Praise for

Considering the dozens of books and zines about biodiesel, the hundreds of workshops held, and the thousands of pages of online discussion, it is a wonder this is the first time our collective knowledge about backyard biodiesel production has been gathered into an easy to read, how-to resource. Piedmont is a commercial operation with a homebrewer heritage. Bob and Lyle bring wisdom born of experience that few else in the industry can, written with grace and humor. This is a must read for the biodiesel nerd. It will sit on my bookshelf between my signed copy of Martin Mittelbach and my faded girlMark zine.

—MATT RUDOLF, former Executive Director,
Piedmont Biofuels Cooperative and ongoing B100 user

Small biodiesel producers have always been the most creative and enthusiastic problem solvers. This book presents the accumulated experience of two of the field's most experienced and authoritative practitioners. They not only guide you through the steps of making your own fuel, but make you excited to get started.

—JON VAN GERPEN, Chair and Professor,
Biological Engineering, University of Idaho

At last! A biofuel book that is realistic about global supply potential while being hands-on useful to the backyard or garage enthusiast. Though some first-generation biodiesel books were hampered by hype and misinformation, Estill and Armantrout get it right.

—RICHARD HEINBERG, author, *Afterburn*

Lyle Estill and Bob Armantrout are some of the absolute best minds in the business when it comes to making biodiesel. Whether it's at a small scale level in your garage or a full-blown commercial biodiesel plant, these guys really know their stuff! Having a book written by these two on how to make biodiesel (and how to avoid all the mistakes along the way) is like having them right beside you. The writing is quick and easy to follow, and the chemistry explanations are great. We highly recommend it as a great book to anyone interested in making their own biodiesel, regardless of the scale.

—GRAYDON BLAIR, President, Utah Biodiesel Supply

While the economic climate may not be ripe for commercial biodiesel production, it is an essential science whose art must not be forgotten purely due to lack of profitability. Having themselves ridden the commercial roller coaster many times, the combined experience of these two authors along with their entertaining rhetoric provide a sound source for safe community-scale biodiesel production to promote clean domestic fuel production and build resilient communities.

—C. DAVID THORNTON, Clemson University Facilities Coordinator for Organics Recycling and Biofuels Programs

BACKYARD BIODIESEL

BACKYARD BIODIESEL

HOW TO BREW
YOUR OWN FUEL

Lyle Estill & Bob Armantrout

new society
PUBLISHERS

Cover design by Diane McIntosh. Fence background: iStock andreusK;
car: iStock rawpixel; beaker: iStock wfenguss

Printed in Canada. First printing April 2015.

New Society Publishers acknowledges the financial support of the Government of
Canada through the Canada Book Fund (CBF) for our publishing activities.

This book is intended to be educational and informative.
It is not intended to serve as a guide. The author and publisher disclaim
all responsibility for any liability, loss or risk that may be associated with
the application of any of the contents of this book.

Inquiries regarding requests to reprint all or part of *Backyard Biodiesel*
should be addressed to New Society Publishers at the address below.
To order directly from the publishers, please call toll-free (North America)
1-800-567-6772, or order online at www.newsociety.com

Any other inquiries can be directed by mail to:

New Society Publishers
P.O. Box 189, Gabriola Island, BC V0R 1X0, Canada
(250) 247-9737

LIBRARY AND ARCHIVES CANADA CATALOGUING IN PUBLICATION

Estill, Lyle, author
Backyard biodiesel : how to brew your own fuel / Lyle Estill
& Bob Armantrout.

Includes index. Issued in print and electronic formats.
ISBN 978-0-86571-785-5 (pbk.).—ISBN 978-1-55092-591-3 (ebook)

1. Biodiesel fuels. I. Armantrout, Bob, author II. Title.

TP359.B46E87 2015   665.37   C2015-901161-2
                                           C2015-901162-0

New Society Publishers' mission is to publish books that contribute in fundamental
ways to building an ecologically sustainable and just society, and to do so with the
least possible impact on the environment, in a manner that models this vision. We
are committed to doing this not just through education, but through action. The
interior pages of our bound books are printed on Forest Stewardship Council®-
registered acid-free paper that is 100% post-consumer recycled (100% old growth
forest-free), processed chlorine-free, and printed with vegetable-based, low-VOC
inks, with covers produced using FSC®-registered stock. New Society also works to
reduce its carbon footprint, and purchases carbon offsets based on an annual audit
to ensure a carbon neutral footprint. For further information, or to browse our
full list of books and purchase securely, visit our website at: www.newsociety.com

For Tami and Camille,
two great women who have been wedded to biodiesel
for better and for worse....

Contents

Acknowledgments

Lyle Estill:
My journey into fuel making in the backyard began with Gary Thompson. He's my ideation wingman who has accidentally approved of my flops, incorrectly vetoed my best ideas and informed my thinking about all things over the years. Most of my backyard brewing was done with Rachel Burton and Leif Forer, who founded Piedmont Biofuels with me before heading off to greener pastures. I'm deeply grateful to them both.

I'm also indebted to the dozens of interns and big-brained people who have passed through our project over the years—taking backyard brewing to great heights. This includes, but is not limited to, Chris Jude, Matt Rudolf, David Thornton, Greg Austic, Tim Angert, Spencer Johnson, Caleb Daniels and all of the Tuesday Night Fuel Makers who continued to carry the torch while I was off in commercial production.

I'm also indebted to all those who have served on Piedmont's Board of Directors—too numerous to list, with some inspirational leadership from Don Mueller, John Bonitz, Brian Gullette, Carol Peppe Hewitt and John Hollingsworth. Without our dedicated co-op members we would not know how to fuel a community as we do now. I appreciate them sticking with us, through good times and bad. The same is true of our restaurant partners, who make everything possible, and those who house our dispensers.

George Seiz, Larry Larson, Eric Henry and Jeff Barney are all candidates for elevation in my book—for putting up with us and for helping us to keep the fuel flowing to the people. Also thanks to Moya Hallstein and Kate Renner for the thousands of gallons of feedstock they delivered to the gate.

I think I also need to acknowledge those industry inspirations that have kept us motivated over time, including Kumar Plocher, Jennifer Radtke, Jon Van Gerpen and Joe Jobe. These folks probably don't know it, but their leadership, perseverance and guidance have had a tremendous impact on the ability of our project to endure.

I also need to thank my brothers, Jim and Glen Estill, for their patience with me, and for accompanying me in my exploration of all things biodiesel. I need to add Tami Schwerin to that list. She is an awesome muse, gadfly, B100 user, copy editor and sympathizer all in one. Dedicating the book to her hardly seems like enough.

I have a deep gratitude to the cast and crew of Piedmont Biofuels—the commercial production crowd. If not for the steady endurance of Link Shumaker, Ray Robinson, Eric Goldston, Katy Horton, Sam Parker, McCayne Miller and many others, I never would have had the opportunity to write this book. Special thanks to Paul Eudy not only for steering the Piedmont ship, and for creating my "shake and bake" reactants that I occasionally use on stage, but also for being available to answer my technical questions—especially during disputes with Bob.

Of course, I need to thank Ingrid Witvoet and the folks at New Society Publishers who have been shepherding my content to market since 2005.

Finally I need to thank Bob and Camille Armantrout. Much of my thinking about all things biodiesel has been shaped in their living rooms, or in their gardens or on their back porch. They have anchored our project with a wisdom and experience for which I am eternally grateful.

— Lyle Estill
October 2014

Bob Armantrout:

I was introduced to biodiesel on Maui in September of 2000 when I met Shaun Stenshol of Maui Recycling Service. Shaun continues to run his recycling fleet, as well as his rental cars, on B100. His values and perseverance continue to inspire me (bio-beetle.com).

I was introduced to the world of petroleum marketing through my association with Alec McBarnet of Maui Oil Company, whose patience and determination taught me a lot about how to succeed in the fuel distribution world.

My work with Russ Teall, Charles Fiedler and the rest of the gang at Biodiesel Industries gave me the opportunity to manage a small plant, and to learn good lab practices through hands-on operation of a gas chromatograph.

I am thrilled to have made a stage dive into Chatham County, North Carolina, with my wife, Camille, and landed among a group of fuel makers, farmers, artisans and educators. Upon my arrival, Andy McMahan engaged me to help him develop and deliver one of the first biofuels curriculums at a community college in the United States. Lyle Estill, and all the folks at Piedmont Biofuels, have kept us engaged in our new community as well as kept Blanche, my 1987 Mercedes TD wagon, filled up with the good stuff.

And lastly, I'd like to acknowledge those of you who comprise the small world of community-scale biodiesel. It is truly the people associated with it who make this fuel so darn ad-dictive. The folks who have worked hard to plan and deliver conferences that allow us all a moment to get together and trade the stories of our successes, failures and spills are to be applauded. The Sustainable Biodiesel Summit led the way, and was joined by the Collective Biodiesel Conference that is still rocking it after all these years.

Thanks to Graydon Blair, Brian Roberts, Jon Meuser, Dara Lor, John Bush and Todd Hill for selecting Pittsboro as the home of the 2014 Collective Biodiesel Conference.

What a long strange trip it's been!

— Bob Armantrout
October 2014

Warning and Disclaimer

Biodiesel production requires the use of hazardous materials like methanol, and caustic materials like potassium hydroxide or sodium hydroxide. These materials can cause serious injury. You are responsible for deciding whether you want to make your own fuel, what the legal and permitting requirements are in your area and how you decide to handle and store these materials.

That being said, the gasoline in your garage for your lawn mower and the drain cleaner under your sink are also dangerous flammable and caustic materials. The ingredients you use to make biodiesel are no more hazardous, other than you may have them on your property in larger quantities.

The information offered here is to help guide you in your own fuel-making journey. It is not intended to be all-inclusive. It's some helpful pointers from a couple of guys who have spent a fair part of their professional lives making biodiesel. Please take the time and make the effort to learn how to make this fuel in a safe and environmentally friendly manner.

Bob and I provide no warranty, expressed or implied, as to the completeness, accuracy or reliability of the information provided. We own the errors and omissions—some of which are intentional.

Introduction

Poor old biodiesel is badly misunderstood.

Some people think that it can save the world. It can't. Others think that biodiesel is the answer to really big, unfathomably complex problems, like peak oil, or America's dependence on foreign oil. It's not.

Many people think that biodiesel is sexy. Sometimes it is. Others think it is evil. It can be. Still others are confused about what role biodiesel can play in America's energy mix. As a fuel, it is both dismissed and embraced by environmentalists, survivalists, politicians and everyone in between.

Biodiesel is made from fats, oils and greases. The word "grease" typically indicates the liquid version of "fat."

Its problem is a function of scale. It is entirely possible to connect with a local restaurant to collect some used fryer oil to make enough fuel to power your family, or your neighborhood or your small town with biodiesel. That's the sexy part.

The problem lies with the human animal's desire for growth. Or more simply, the problem lies with greed. When you successfully make your first gallon of fuel, it's exciting, and it makes you want to make a hundred gallons more. When you have made your first hundred gallons, it makes you want to make a million gallons. And when you do that, you want to make a hundred million gallons more.

Since you will rapidly run out of used fats, oils and greases, you will need another feedstock to make enough fuel. Let's do this: burn down the rainforest in Malaysia and plant oil palm trees all in a row. Crush the palm seeds into oil, load the oil onto a supertanker and send it to Seattle, where there is a one-hundred-million-gallon biodiesel plant in the harbor. Spin the palm oil into fuel, put it on a train, send the train to North Carolina, put the product on a truck and take it to the airport to burn in their fleet. When they have burned enough of it, let's give them an award for being "green."

That's how biodiesel can be evil. Stop the madness.

And the reverse is also true.

I work at Piedmont Biofuels. We are a community-scale biodiesel plant with a one-million-gallon per year capacity. There was a time when the American dollar fell off a cliff, and the euro was still strong. That was when Europeans were taking shopping trips to New York, and hotdog vendors were accepting euros on the street corner.

During that period, when America was at a discount, the Europeans came over and bought up all of the poultry fat in the southeastern United States. They then approached eleven biodiesel plants (including Piedmont) and essentially said, "Now that we own all of the feedstock, how would you like to work for us?"

We all agreed. They rented a three-million-gallon tank in the Savannah harbor with a floating lid, and we all shipped product to them. Piedmont turned a tanker load of fuel every other day, and the sum total of our endeavor would raise the lid on that tank by one inch.

When they had aggregated enough fuel, they would ship a supertanker from Savannah to Rotterdam, blend our product into the European fuel system and deliver it to the street corner in London so that the fellow filling up his diesel vehicle could call himself "green."

Again, stop the madness.

Biodiesel can be evil. It can also be a good thing. Perhaps Kent Bullard said it best. He used to procure biodiesel to power his national park, and as a champion of the product, he liked to say, "Biodiesel is one tool in the sustainability toolbox."

He's right about that. If you can lay your hands on some waste feedstock, you can convert it into fuel to provide motive power, and that can be a good thing. Especially if you can stay small.

Everyone admires the small-scale producer—especially the backyard brewer who is merely meeting his or her fuel needs, and clobbering the price at the pump. But again, we need to be careful.

Making your own fuel is like heating with wood. It takes work. Lots of work.

People gathered around the warm radiant heat of a woodstove tend to romanticize "heating with wood." If you live in the woods, like we do, firewood is "free."

What typically gets left out of the equation is the skidding, chainsawing, splitting and curing that goes into a decent piece of firewood. As a heat source, it is deeply satisfying, and expensive.

The same is true of backyard biodiesel. Making fuel is hot, heavy, smelly, dangerous work, and getting a tankful can be a chore. But once it is done, driving down the road on fuel you have made yourself is a liberating, exhilarating and wonderful feeling.

Once the work is done, it's easy to enumerate biodiesel's charms: everything from "Made in America" to "No War Required." People who drive around on B100 (100% biodiesel) are largely off the petroleum grid—free of dead pelicans on the evening news, exempt from tithing to Halliburton.

If you want to make your own fuel, this book can help lighten the load. It can help you get the recipe right, and it offers lots of stories and insights that will save you grief, and work.

Bob and I have a lot of experience making biodiesel. I was a backyard brewer when Bob was running a commercial plant. I was running Piedmont Biofuels when Bob started teaching biodiesel at Central Carolina Community College. We are intimate with this fuel. Our talents combined have already made most of the mistakes that can be made when making fuel, or collecting oil or distributing product. Our intent is to lay those bare to save you thousands of dollars, and massive amounts of work.

I should note that this book occasionally bumps up and down between backyard fuel making and commercial production. We have tried to keep a fence around commercial production, intentionally attempting to leave it out. But to do so completely would give an incomplete story. Things like ASTM D6751 is the commercial specification for biodiesel, and its components serve as a guide for backyard fuel quality. And like it or not, backyard fuel makers are part of the biodiesel industry.

Tensions between backyard brewers, community-scale biodiesel plants and giant commercial concerns have lessened considerably over the past decade. There is now an acceptance of small producers by the National Biodiesel Board, and the industry's flagship periodical, *Biodiesel Magazine*, has increasingly covered small producers over the years. While it might be a begrudging acceptance, there is an understanding that the people on the ground making their own fuel are the ones that create the buzz for the industry in general. The popularization of biodiesel as a fuel is due in large part to the accidental public relations work done by those making fuel in the backyard.

Hopefully you will enjoy the pithy wisdom, or the practical shortcuts or the tales of sheer stupidity contained herein.

Of the two of us, I'm the storyteller, and Bob is the operations guy. I'm good at broad outlines, and thirty-thousand-foot views, and Bob is good at the details and the technical stuff.

We think it will be easy for you to tell our two voices apart. I do the style; Bob does the substance. When the writing is facetious, personal and over-endowed with self-importance, that's Lyle. When the writing is straightforward and technical, that's Bob.

A note of caution: When making your own fuel, it is easy to get lost in technical details. And technical details are important. But after all the titrations, and BTU measurements and pump preferences have faded—after all the opinions, and false claims and misinformation have been forgotten, the empowerment that accompanies making fuel from scratch is amazing.

When we get into the nuts and bolts of backyard biodiesel production, we tend to refer to Amazon as a baseline lowest price. Clearly it would be better for you, and your local economy and for biodiesel in general, to spend your dollars at your locally owned hardware store, or at one of the biodiesel parts retailers mentioned in this book.

In a world of top-down energy choices, where you need to inherit an oil well in order to participate, biodiesel is empowering.

Perhaps it is the empowerment that counts. That's the point of this book. To empower you.

Literature Review

There is a canon of biodiesel literature. We can debate where it begins and ends. And we can argue about what texts are worthy of inclusion, but there is no doubt that many books have influenced and shaped the world of biodiesel.

For me, backyard biodiesel begins *From the Fryer to the Fuel Tank*, which Josh Tickell self-published in 2000. People love to bash on Tickell's work. Some say it lacks the proper safety component. Some say it was not updated regularly enough. But I think Josh Tickell created backyard biodiesel in America.

Josh is a complicated character. He's part pioneer, part homebrewer and part community builder. Just as the home-brew community was turning his back on him, he published *Biodiesel America*, with the grandiose subtitle, *How to Achieve Energy Security, Free America from Middle-East Oil Dependence and Make Money Growing Fuel.*

I thought it was great book, and I gave it a good review on Energy Blog (biofuels.coop/biodiesel-america). It was published in 2006 with the support of the National Biodiesel

Board, which fed the fire of Tickell detractors. Those who want to characterize him as an industry shill overlook his vast contributions to biodiesel in this country.

As someone who does not like to be "out-self-promoted," I have to tip my hat to Josh. Both of his biodiesel books are available on Amazon, but have disappeared from his website. He seems to have moved on to making movies, doing speaking gigs and talking about millennials.

For some, the cornerstone of biodiesel literature is Martin Mittelbach and Claudia Remschmidt's *Biodiesel: A Comprehensive Handbook*, published by John Wiley and Sons in 2004. They have an academic approach, with a European perspective, that is easy to dismiss as irrelevant to the American homebrewing scene, but it remains a book we all turn to from time to time.

Another important milestone in biodiesel literature was Shaine Tyson's *Biodiesel Handling and Use Guidelines*, published by the Department of Energy in 2004. The fourth edition, published in 2009, can be downloaded as a PDF (nrel.gov /vehiclesandfuels/pdfs/43672.pdf).

My first book, *Biodiesel Power: The Passion, the People, and the Politics of the Next Renewable Fuel*, was published by New Society Publishers in 2005. The National Biodiesel Board banned its sale at the 2006 national conference. That hurt my feelings. But it didn't hurt sales. It went through multiple printings and became a cult classic for homebrewers everywhere. It's not a how-to book. It's simply stories of the industry, and of the journey we took at Piedmont Biofuels from backyard production to a community-scale project.

Biodiesel Power came out about the same time as Greg Pahl's *Biodiesel: Growing a New Energy Economy*, published by Chelsea Green in 2005. I was terrified by its presence in the market, but upon reading it, I realized it was a very different text. Pahl took a journalistic survey approach to the industry. His work is very different from my first-person angle on the subject. I gave Greg's book a good review on Energy Blog (bio fuels.coop/greg-pahls-new-book).

A literature review of the biodiesel canon would not be complete without Jon Van Gerpen and his two significant contributions. *The Biodiesel Handbook*, co-authored with Gerhard Knothe and Jurgen Krahl, was published by the American Oleo Chemical Society's AOCS Press in 2005. He also wrote *Building a Successful Biodiesel Business: Technology Considerations, Developing the Business, Analytical Methodologies* with co-authors Rudy Pruszko, Davis Clements and Brent Shanks and self-published it in 2005. Both books are academic in nature. Both are expensive, helpful and obscure—although still available on Amazon.

Van Gerpen's contribution to backyard brewing goes well beyond his publications. As a professor at Iowa State and the University of Idaho, he has taught and influenced many backyard brewers and commercial producers nationwide. You could say that Jon Van Gerpen is the king of biodiesel in this country. Or at least the crown prince.

In 2006, William Kemp's wide-sweeping book *Biodiesel Basics and Beyond* was published by Aztext press and distributed by New Society Publishers. It is filled with compelling stories, and photographs, with a decidedly Canadian angle.

I need to mention the self-published zines that have been critically important in moving the biodiesel movement along. First and foremost is Girl Mark Alovert's *Biodiesel Homebrewer's Guide*, which went through many revisions and editions before Girl Mark vanished from the biodiesel community. My copy, the tenth edition, was stapled together in 2003.

Perhaps more germane is Matt Steiman's *Biodiesel Safety and Best Management Practices for Small-Scale Noncommercial Use and Production*, published by Pennsylvania State's College of Agricultural Science in 2007. A PDF is available online (pubs.cas.psu.edu/FreePubs/pdfs/agrs103.pdf).

Also important is Jennifer Radtke's self-published zine, *Not a Gas Station*, which recounts the terrific story of Biofuels ·Oasis in Berkeley, California. Published in 2006, it can be purchased online (biofueloasis.com/faq/jrs-book).

Another valuable contribution to the backyard biodiesel movement is *BiodieselSMARTER*, a now defunct magazine published by Frankie Abralind. Homebrewers filled its issues with stories, and it left an indelible mark on the community. It appeared in twelve quarterly issues from 2007 through 2009, the last issue of which was the B100 Map that chronicled all of the grassroots biodiesel dispensing stations in the land.

BiodieselSMARTER gets mentioned in Jon Starbuck and Gavin Harper's wonderful book *Run Your Diesel Vehicle on Biofuels*, published by McGraw Hill in 2009. They hail from the other side of the pond, in England, and they tapped into the US scene in 2007 at the National Biodiesel Conference. *Run Your Diesel Vehicle on Biofuels* was cleverly brought to you by the letter B, and the number 100.

BBI International—a conglomerate with a handful of titles and exhibitions—began publishing *Biodiesel Magazine* in 2004. It became a monthly in 2006, and fell back to bi-monthly in 2012. What began as a pass-through industry rag started taking editorial shape (including stories of small-scale co-ops and producers) under the leadership of Ron Kotrba, beginning in 2009. The appearance of an article entitled "The Resilient Community Producers" in 2010 marked a watershed moment for small-scale biodiesel in America (biodieselmagazine.com /articles/4290/the-resilient-community-producers).

A lot of self-published work, including zines, have contributed mightily to the development of both the biodiesel industry and to backyard biodiesel—but beware of self-published biodiesel books on Amazon by folks you have never heard of. I bought a handful in preparation for this literature review and was sorely disappointed. Their errors and omissions outweigh whatever credit they might deserve for economic opportunism.

These are the books that fill the shelves of everyone in the biodiesel industry. Whether it's a commercial producer stuck with a stubborn emulsion who reaches for their Mittelbach, or a homebrewer thinking about entering the business while studying their Van Gerpen, this is the literary tradition from which we hail.

Backyard Biodiesel intends to help you meet your fuel needs. Our hope is that this book finds its way into your garage, has its spine broken by an old brick and ends up with grease stains penetrating the pages.

Oil and Biodiesel
Chemistry Primer

Before you embark on your backyard biodiesel journey, a basic understanding of the chemistry of fats, oils and biodiesel can be helpful. The chemistry behind fats and oils is fairly straightforward, and fortunately biodiesel chemistry is closely related.

Molecular Structure of Fats and Oils

Fats and oils are a type of material broadly defined as lipids. A lipid is any fat-soluble molecule, that is, one that will dissolve in fat. Lipids include not only animal fats and vegetable oils, but also molecules like vitamin E (tocopherol), sterols (like cholesterol), waxes (long-chain fatty acids) and gums (phosphatides).

Typically "fats" are solid at room temperature and "oils" are liquid. "Fat" is often the term used for lipids rendered from animal products, and "oils" is used for the lipids produced from vegetable products.

All fats and oils primarily consist of a molecule called a **triglyceride**, the combination of a glycerol (aka glycerin) molecule and three **fatty acid** chains. It looks like this:

The glycerin "backbone" is to the left in this diagram, with three fatty acid chains attached off to the right. The type of fatty acid is determined by the chain length as well as the number of double bonds (aka saturation level) in the chain. In this diagram, the top chain has 16 carbon atoms with no double bonds, the middle chain has 18 carbons with 1 double bond, and the bottom one has 18 carbons with 3 double bonds. They are (in the same order) palmitic acid, oleic acid and linolenic acid. These acids can also be described numerically, with their carbon chain length and saturation level like this—16:0 (palmitic), 18:1 (oleic) and 18:3 (linolenic).

Fatty acids' molecular structure can also be expressed graphically like this:

This is oleic acid again, the most common fatty acid in fats and oils. Note the single double bond in the middle of the chain, reducing the number of potential hydrogen atoms by two. Due to the presence of a single double bond, it is monounsaturated.

The saturation level of the fatty acid also affects the properties of the biodiesel produced from it. A fatty acid is considered saturated when there are no double bonds between carbon atoms, and therefore it can hold no more hydrogen atoms. Fatty acids with one double bond are monounsaturated and those with more than one double bond are polyunsaturated. The more saturated the fat, the higher the temperature at which the fuel will gel. And the more saturated the fat, the better the oxidative stability.

Used fryer oil is typically a mix of vegetable oils—soybean, palm, canola, etc.—and usually doesn't contain too many saturated triglycerides. It's always a good idea to ask the restaurant whose grease you are collecting to show you the box that the oil came in so you know what you're dealing with.

Reactions Involving Triglycerides

Triglycerides commonly undergo a number of reactions. Understanding the nature of these reactions can help you learn to make better fuel easier. It may be a bit confusing at first, but stick with it and your increased knowledge will pay off.

Hydrolysis of fats or oils readily occurs when water is present. This reaction causes one (or more) of the fatty acids to detach from the triglyceride. The products of this reaction are one glycerin molecule and three free fatty acid molecules. If two fatty acids remain attached to the glycerin backbone, a **diglyceride** is the result. Only one fatty acid chain bound to the glycerin backbone is known as a **monoglyceride**. Hydrolysis of oil is the major contributor to the degradation of used cooking oil through the formation of free fatty acids. Water is to be avoided whenever possible.

Oxidation of fats and oils is degradation resulting from the presence of oxygen. At room temperature, this is a slow reaction compared to hydrolysis, but at frying temperatures, oxidation speeds up. Auto-oxidation occurs at ambient (room) temperature and is responsible for oil turning rancid, resulting in off smells and flavors. Exposure to light increases the rate of oxidation, so keeping your oil out of direct light will reduce its degradation.

Polymerization of fats and oils occurs when fatty acid chains are broken off the triglyceride and then recombine with other free fatty acids (**FFA**) to form longer fatty acid chains or when triglyceride molecules combine to form large molecules. Polymerization increases the thickness (viscosity) of the oil and is responsible for the unremovable "goo" that coats the outside of oil collection containers. Polymerization occurs fairly slowly, and as such, doesn't come into play much in backyard biodiesel production. If you can keep hydrolysis and oxidation at bay, polymerization won't be a problem.

Now let's turn our attention to the biodiesel side of the equation. The good news is that the reactions that degrade oil are the same ones that will ruin your biodiesel, and the way to avoid them is the same too.

Molecular Structure of Biodiesel

Biodiesel is known chemically as a methyl ester. An ester is the result of a reaction between an alcohol and an acid. Biodiesel is formed from a reaction between methanol (an alcohol) and a free fatty acid. Its molecular structure looks like this:

$$H-\underset{\underset{H}{|}}{\overset{\overset{H}{|}}{C}}-\underset{\underset{H}{|}}{\overset{\overset{H}{|}}{C}}-\underset{\underset{H}{|}}{\overset{\overset{H}{|}}{C}}-\underset{\underset{H}{|}}{\overset{\overset{H}{|}}{C}}-\underset{\underset{H}{|}}{\overset{\overset{H}{|}}{C}}-\underset{\underset{H}{|}}{\overset{\overset{H}{|}}{C}}-\underset{\underset{H}{|}}{\overset{\overset{H}{|}}{C}}-\underset{\underset{H}{|}}{\overset{\overset{H}{|}}{C}}-\overset{\overset{H}{|}}{C}=\overset{\overset{H}{|}}{C}-\underset{\underset{H}{|}}{\overset{\overset{H}{|}}{C}}-\underset{\underset{H}{|}}{\overset{\overset{H}{|}}{C}}-\underset{\underset{H}{|}}{\overset{\overset{H}{|}}{C}}-\underset{\underset{H}{|}}{\overset{\overset{H}{|}}{C}}-\underset{\underset{H}{|}}{\overset{\overset{H}{|}}{C}}-\underset{\underset{H}{|}}{\overset{\overset{H}{|}}{C}}-\underset{\underset{H}{|}}{\overset{\overset{H}{|}}{C}}-\overset{\overset{O}{\|}}{C}-O-\underset{\underset{H}{|}}{\overset{\overset{H}{|}}{C}}-H$$

Note how similar this structure is to oleic acid. The OH group on the end has been replaced by a CH3 (methyl) group.

The type of fatty acid that comprises the fuel dictates the physical properties of the fuel produced. The more saturated the fatty acid, the higher the gel point of the fuel, but the more shelf stable it is—just like with the original oil.

Reactions Involving Biodiesel

Remember that the same reactions that make your oil go bad will make your biodiesel go bad as well.

Hydrolysis, oxidation and polymerization can all occur in improperly made or stored methyl esters by the same mechanisms involved with your feedstock.

In addition to those reactions, a few more come into play during different phases of biodiesel production that you should be aware of and get to know.

Transesterification is the primary reaction we use to make biodiesel. It is the creation of three molecules of methyl ester from the reaction of one triglyceride molecule and three methanol molecules in the presence of a catalyst. A catalyst is a material that increases the rate of a reaction without being changed or consumed. In our backyard fuel making, we'll use potassium hydroxide (KOH) or sodium hydroxide (NaOH) as our catalyst for the transesterification reaction. Three molecules of methanol are required for each triglyceride in order to produce the three molecules of biodiesel by attaching one methyl group to the end of each of the three fatty acid chains present. The transesterification reaction looks like this:

$$(1) \text{triglyceride} + (3) \text{methanol} \underset{}{\overset{\text{catalyst}}{\rightleftharpoons}} (3) \text{methyl ester} + (1) \text{ glycerol}$$

Note that the arrows point both ways, indicating that this reaction is capable of going both ways. We'll discuss this more a bit later in the Recipes section.

It is called transesterification because one ester, a triglyceride made of the alcohol glycerol and three free fatty acids, is converted (transformed) to a different ester—biodiesel, a methyl ester comprised of the alcohol methanol and one free fatty acid chain.

Esterification is the reaction where a free fatty acid is converted to biodiesel. It typically uses sulfuric acid as the catalyst rather than a base like KOH. In the esterification reaction, a methyl group is bonded onto the FFA, and a molecule of water is formed as a by-product. The esterification reaction looks like this:

(1)free fatty acid + (1)methanol $\underset{}{\overset{\text{catalyst}}{\rightleftharpoons}}$ (1)methyl ester + (1)water

Esterification is a useful advanced technique for getting better yield from poor-quality oil, as we shall see.

We use the term **conversion** to describe the extent to which our feedstock has been turned into biodiesel, and not left as a mono-, di- or triglyceride, an FFA or turned into a soap.

Saponification is the "soap-making" reaction that biodiesel producers want to minimize, as it leads to reduced yields. It occurs when a free fatty acid meets a base catalyst and the metal (potassium or sodium) is bonded to the end of the FFA, creating a salt of the fatty acid. A salt is the product of an acid-base reaction, and in this case, we have combined a fatty acid with the base KOH or NaOH. When the salt contains a fatty acid, we call it a soap. The saponification reaction looks like this:

(1)free fatty acid + (1)KOH or NaOH $\underset{}{\overset{\text{catalyst}}{\rightleftharpoons}}$ (1)KOH or NaOH salt (soap) + (1)water

Note that additional water is a by-product of the saponification reaction. Water contributes to the hydrolysis reaction that degrades oil or fuel. Also note that the saponification reaction proceeds much faster than transesterification; therefore, any FFA in your feedstock will be turned into soap (and water) not biodiesel.

Soap is an **emulsifier**. An emulsifier is a molecule that is able to bond to both polar and non-polar compounds—like water (polar) and biodiesel (non-polar). This is why soap is able to get grease off your hands when washing with water. The polar water and the non-polar grease have no ability to bond until the emulsifying soap is used. The soap binds to the oil and the water, and they all go down the drain together. Emulsions are to be avoided when making biodiesel.

Polarity of molecules is a continuum, not an either-or. Materials are said to be relatively polar or relatively non-polar. A basic understanding of the polar nature of common biodiesel production materials follows.

Relatively polar materials used or created in biodiesel production include water, methanol, glycerin, base catalysts and salts (soaps).

Relatively non-polar materials used or created in biodiesel production include fats and oil, and biodiesel.

This means that when we produce our transesterification reaction, and the two liquid "phases" separate, the relatively less dense non-polar phase on top will contain the biodiesel, any unreacted oil and small amounts of glycerin, methanol, water, catalyst and soap. The heavier polar phase will contain glycerin, methanol, base catalyst, soaps, water and small amounts of biodiesel and unreacted oil.

The fact that the "contaminants" of biodiesel glycerin, water, methanol and soaps are all relatively polar accounts for the efficacy of water washing our fuel to remove them. The basic chemistry concept of "Like dissolves like" works to our advantage.

Reaction Kinetics

Reaction kinetics describe the rate with which a chemical reaction occurs and the factors that affect the rate. In the transesterification reaction, three factors that greatly affect the rate of the reaction are:

TIME — TEMPERATURE — TURBULENCE

The more you increase one factor (to a point), the more you can decrease the other two and still push the reaction forward. Since non-polar oil and polar methanol and caustic don't want to readily mix, increasing the turbulence of mixing will help ensure that methanol is available at the point when the triglyceride is being deconstructed by the catalyst. It is also for this reason that we typically use twice the amount of methanol required to "push" the reaction in the direction we want it to go, as well as to reduce soap production. Dependent on how turbulent your mixing is, you can reduce the mixing time and still get complete conversion of triglycerides into methyl esters.

The higher the temperature of your reaction, the faster it will proceed as well. Keep in mind that the boiling point of methanol (at sea level) is 148.5°F (64.7°C). This means that if your reaction gets hotter than that, you risk boiling off some of your methanol, thereby wasting it and creating additional fumes. Additionally, the transesterification reaction is slightly exothermic (generates its own heat), and so keeping your feedstock at about 131°–140°F (55°–60°C) is ideal as you begin your reaction.

Typical reaction times are 90 minutes or more, depending on what type of mixing your reactor uses.

Now that you've been armed with some basic organic chemistry, let's get on to the business of preparing our materials to make some fuel.

Gathering Ingredients

Note: We'll cover the handling, storage and safety concerns surrounding our ingredients in the Safety chapter. Make sure you've read and understood the safe way to handle and store dangerous materials before bringing them onto your property.

Oil Quality

The quality of your oil feedstock is a major determinant of the ease in converting it into quality biodiesel. The major variables that come into play are the levels of moisture (water), FFA, saturated fats and particulates.

Eliminating as much water as possible from your oil is crucial towards getting good conversion, low soap production and a good yield. The good news is that it's not very hard to do.

Having low FFA content in your oil makes processing much easier, as soap production will be greatly reduced. The easiest way to do this is to start with low FFA oil. Other possibilities include esterifying your feedstock first, to convert the FFA directly to biodiesel, and then transesterifying the batch to convert the glycerides.

The saturation level of the fatty acids (free or bound to glycerin) will affect your fuel quality as well. Remember, the more saturated the fatty acids, the lower the cloud point of your fuel. This means if you use highly saturated feedstocks, like beef tallow or palm oil, expect the fuel you produce to start to solidify in the mid-60°F (15.5°C) range. This is problematic in much of the country much of the year, and these feedstocks should be avoided if possible.

Particulate matter and other solid trash in the oil is not a big problem to remove through settling, but the more of it you have, the more you'll need to find a way to get rid of it.

So, work hard to get the best oil you can, and don't hesitate to drop sources whose oil is suffering from one of the aforementioned contaminants.

Methanol

Methanol is the alcohol of choice for backyard fuel making. Ethanol is too tricky to work with, due to its extreme affinity for water, and is prohibitively expensive as well. Methanol is both toxic and flammable, so make sure you've read the contents of the Safety chapter before purchasing and handling it.

Methanol is consumed in a 3 to 1 molar ratio to the triglycerides in our feedstock. As mentioned earlier, because we want to "push" the reaction towards the biodiesel side, we typically use twice the amount required, or 6 to 1 molar ratio. This means for each triglyceride molecule, we'll have six methanol molecules available for it to react with. Methanol is expensive, and because we're using twice the amount actually consumed in the reaction, we'll want to recover the excess for reuse in our next batch.

Methanol quality is crucial to good fuel production. The major contaminant in methanol is water, as they are both quite polar and therefore have a high affinity for each other. Make sure the methanol you're using is 99.5% pure or better. This can be purchased online, but it's very expensive ($4.16/gallon if purchased online or US$3.70/gallon if purchased from racing fuel suppliers).

The cost of the methanol per gallon of biodiesel produced (assuming a 90% yield and no recovery) will be about $0.90 at racing fuel pricing. These numbers will be cut by about 40% if you're able to reuse your recovered methanol. Clearly, purchasing methanol online is not the way to go. Most chemical distributors will be happy to sell you methanol in 55-gallon drums.

KOH Versus NaOH

You have a choice of which catalyst you use for your fuel making—KOH or NaOH. There is a tradeoff: KOH is much more forgiving (easier to use successfully), creates a glycerin phase that is liquid at room temperature and is much more environmentally friendly, if you intend to compost or otherwise land apply your glycerin—but it is more expensive.

NaOH is the opposite on all the same factors. It is trickier to work with (more susceptible to producing "glop" if a bit too much is used, produces a glycerin phase that sets up at room temperature and, due to the sodium salt produced, is toxic to most plants. It also doesn't dissolve in methanol as readily as KOH does when preparing your catalyst.

Buying catalyst can be done online, but the shipping costs are high due to it being hazardous material. If you can find a local supplier, you'll be better off. Often you'll be required to have a business license to buy in bulk, and a quick cost analysis may show that it'll be worth your while to pay the cost of having the business license, depending on the amount of fuel you plan on making. Both KOH and NaOH are used for soap making, so look for soap-maker supply stores either online or in your locale.

The cost difference between the two is something you can calculate based on the specifics of your process. KOH, as of this writing, sells for about $3.00/lb. (delivered) from online sellers. NaOH sell for about $2.55/lb. You'll use about 60% more KOH by weight than you do NaOH to process the same oil. For a 50-gallon batch, the cost of catalyst per gallon of finished biodiesel (assuming a 90% yield and 2% FFA) is about $0.30/gal for KOH and about $0.15 for NaOH. Try both and decide which one works best for you, given your process and waste disposal goals. Our preference is KOH, as we feel the benefits of using it outweigh the added cost.

Grease Wars

The bad news for biodiesel is that it is no secret that there is good energy in fats, oils and greases—whether they are virgin or gently used.

Most virgin oils enter the food markets. And food markets consistently outbid the fuel markets in the biodiesel business. That means that, in order to be successful, biodiesel must find a feedstock "anomaly." For homebrewers, this typically means hooking up with a restaurant or two, collecting their used fryer oil and using that as the basis for their fuel-making operations.

Those interested in firing up homebrewing operations should start in the back alley, or with a trip behind the mall. Behind each food service establishment, there will be a grease collection vessel, or two. Or three. Most will be labeled and include the telephone number of the vessel owner. Some belong to giant rendering companies who aggregate vast quantities of fats, oils and greases from animal slaughtering operations to used fryer oil. They tend to have large collection vessels that they service at longer intervals.

Used oil goes into animal feeds, and makeup, and candles and soaps. Biodiesel makers merely bid on the resource for fuel. When collecting used cooking oil from restaurants, you are not "recycling" a resource that was going to the landfill.

You hear that a lot in biodiesel. It sounds virtuous. But it's not true.

At Piedmont Biofuels, we have had a lot of luck with county governments. We have deployed dozens of grease collection vessels at recycling centers, so that when the public arrives to recycle their green glass, plastics and cardboards, they can also dispose of used fats and oils. These gallons, which represent an extremely small fraction of the grease we collect, would truly be "incremental" feedstock that was heading down the drain or to the landfill.

In the back alley, the smaller vessels tend to be owned by the biodiesel businesses. They collect oil more frequently in order to arrest the development of free fatty acids, which do not easily convert into fuel. The biodiesel maker who collects those gets to throw them away. The longer oil sits in a collection vessel, especially in the hot sun or in the presence of water, the more degraded the oil becomes; that is, the more free fatty acids are generated, the less valuable the oil becomes as a feedstock for making biodiesel.

When the biodiesel industry is booming, there is a lot of demand for this back alley resource. It becomes so valuable that people steal it.

When I started making fuel in the backyard in 2002, most of the dumpsters in the Research Triangle Park of North Carolina, where I live, were wide open and overflowing. Most restaurateurs would hear that I wanted their used cooking oil to make fuel for my family, and they would invite me to help myself.

Ten years later, those vessels are now watertight, locked and under contract. In North Carolina, it is a felony to take someone else's grease. From a chain of custody perspective, the grease changes ownership once it leaves the restaurant's fryer and enters the collector's vessel.

The big rendering companies have not fared well with biodiesel. They tend to be huge, consolidated and aged. As a result, they have lost touch with their restaurant customers. Companies that are shipping sixteen million pounds a week tend to be unresponsive to a restaurant's 350-gallon problem.

Since backyarders and community-scale producers can easily skate circles around the giant rendering companies, the renderers tend to take a legalistic approach. In North Carolina, for instance, Valley Proteins worked for years at successfully passing a "Grease Police" law that says in order to collect used cooking oil you must have a properly labeled vehicle. They were unsuccessful in their push for licensure, spot inspections and other arduous regulations designed to put small collectors out of business.

All of which is to say, be careful. Before you start setting up a backyard biodiesel operation, learn the laws in your state. They vary across the country like a patchwork quilt. Don't steal other people's grease.

It's one thing to not steal grease. It is a piece of cake to "steal" accounts. If your buddy owns a restaurant, it is easy to intercept the grease before it even makes it to the collector's vessel.

While grease is everything to the biodiesel producer, it is merely a nuisance for the restaurateur. When you are running a restaurant, you have a lot of challenges. You are more worried about whether or not you have enough endive for the nightly special, or about whether your dishwasher is going to show up for work. When it comes to disposal of your waste cooking oil, you just want a clean space and a vessel with enough headspace to dump your spent product into.

If you are going to deploy vessels as a backyard brewer, keep them and the space clean, and keep them empty. Some restaurants will give you their oil in exchange for the service. Some might expect a small remuneration. Very few are paying for grease disposal anymore.

The oil collection business is simply "Service 101." Make sure the customer can get a human on the phone, and make sure you can react to their needs quickly. When a restaurant has a successful fried-food special, or experiences a heavy weekend of traffic, they occasionally need "emergency" servicing. If you can't get out to collect their grease, have another vessel ready to deploy. All they require is headspace and responsiveness. They have a restaurant to run.

Securing your feedstocks is everything in biodiesel. Occasionally you will need to educate your restaurant partners. Water is not your friend when making biodiesel. And when you bring home chunks and bits of fried food, not only do you not have a use for that material, but you will need a strategy to dispose of those things you bring home that cannot be converted into fuel.

In general, expect 20 percent of the material you collect to be "other." At Piedmont, we refer to this as "schmutz." It tends to be non-toxic and compostable, but paying to compost materials can get expensive.

Once again the issue of scale enters into the calculations. In general, the best way to think about fuel production is that a gallon of grease is equal to a gallon of fuel. In the process of making it, you will throw 20 percent other reactants into the process, and you will be left with 20 percent as co-products.

But when doing this calculation, remember that you will often need to dispose of about 20 percent of what you bring home.

So if you want to make 80 gallons a month to meet your fuel needs, you will need to bring home 100 gallons from your grease collection partners. Here is where scale comes in. If your backyard compost heap can easily absorb 20 gallons a month, you are fine. But be careful. Most backyard compost heaps are performing aerobic digestion. Twenty gallons of additional greasy water can turn them into goo and cause them to go anaerobic. Compost heaps that are too wet and turn anaerobic tend to stink. And greasy material is not generally a welcome addition to your compost pile.

The idea is to have a strategy before you begin collecting feedstock. You don't want to put schmutz into a septic system. Most municipal wastewater treatment systems don't want it anywhere near their sewer systems. If you simply dump it on the land, it will tend to kill any vegetation growing there. Grease has a remarkable ability to spread very thin across all surfaces. It will eliminate plant respiration and cause grass, or anything it touches it to turn black.

In most jurisdictions, it is illegal to land apply schmutz. That is largely because of a fear that it will find its way into waterways during rain events.

It's important to note that pollution is also a function of scale. Too much of anything can destroy a creek—whether it is motor oil or mother's milk.

Chances are your local pig farmer will be delighted to take all of your schmutz off your hands. The point is to develop a strategy in advance.

Once you have figured out how to collect used oil, and you have developed a disposal strategy for schmutz, and once you are meeting your fuel requirements, you will have the fantasy of scaling up—so that you can make a living by making fuel.

Again, all biodiesel in America is predicated on feedstock anomalies. These vary over time and space. In Oregon, for instance, the state contract for biodiesel procurement was de-linked from petroleum diesel pricing and instead tied to the market price of canola oil. As a result, Oregon has a robust canola growing industry. And a vibrant seed-crushing industry. And the biodiesel industry is thriving here—on fuel made from virgin oil. This is a policy anomaly.

There was a period of time in Texas when the cotton growers could not extract proper value from their cottonseed production. For a time, biodiesel came to the rescue, making fuel from cottonseed oil. As soon as the cotton growers figured out how to get their oil to global foods markets, the biodiesel-from-cottonseed-oil anomaly collapsed for biodiesel.

Hawaii enjoys a robust biodiesel industry. But Hawaii itself is an anomaly—where most of the inputs to their economy are imported, and most of their waste products are exported.

Many of the giant biodiesel producers in the United States are vertically integrated with soybean growers. By contracting with soy farmers, and owning the storage, drying and crushing infrastructure, they are able to secure a steady source of feed-stock—although they also sell oil into food markets when food commands a more advantageous price than fuel.

Most community-scale biodiesel producers have created their own anomalies by offering a superior service compared to the existing rendering companies.

At Piedmont Biofuels, we successfully linked those who drive around on our fuel to our restaurants through our Partners in Sustainability program. We are a small cooperative of about 400 members. Our members are driving around on 100 percent biodiesel—but they are also "eaters." We let them know who our grease collection partners are, through our newsletters, our Green Dining Guide and Grease Appreciation nights.

When our members are thinking of dining out in Raleigh, and they are in the mood for Asian cuisine, they can hop onto our mobile app and chose a restaurant based on whether or not we collect its used cooking oil. When new restaurants sign up with Piedmont, we can guarantee them new faces—our members, who drive around on B100, and often leave business cards when they pay the bill. The cards read: "Compliments to the chef, I got here tonight on fuel made from your used cooking oil. Thank you."

With our Partners in Sustainability program, we have created our own feedstock anomaly.

If you are embarking on backyard fuel production, that's simple. Find a source for grease. Elbow your way into a crowded market and pick up the feedstocks you need.

Provide good service, do it legally, keep things clean, and you too can be an actor in the "Grease Wars" that go on in most back alleys in America.

Basic Recipes

There are many ways to turn waste cooking oil into biodiesel. The method you choose will depend on the quality of your collected feedstock, the time you want to spend on fuel making, the type of equipment you're using and how important yield is to you. Most of us start off as simple as possible, and then move on to more advanced techniques as our skills and desires develop.

The recipes included here a very good starting point—get started with these and modify as you find your way. Many longtime fuel makers still use the most basic single-stage re-action, along with water washing and bubble drying, to be the simplest and most time- and cost-effective method.

Because feedstock will usually vary in free fatty acid (FFA) content for each batch, we need to adjust our recipe in order to get good results. The more FFA in the feedstock, the more acidic it is. The increased acid neutralizes some of the basic catalyst and produces soap via saponification. In order to push our reaction to completion, we need to increase the amount of base catalyst to make up for the loss through saponification.

Titration to Determine FFA Content

In order to determine the amount of FFA present in each batch of feedstock, you'll need to do a **titration**. In this case, the titration is a method to quantify the amount of FFA present. You can buy the tools and ingredients to perform the titration online. I've itemized them in the Backyard Analytics chapter.

There are two basic methods for titrating oil, the homebrew method and the one commercial producers use. The homebrew titration results in "milliliters of titrant" used, whereas most other folks like to express FFA as a percent of oil weight. I suggest you use the latter process and use the "percent FFA" approach.

Here's the titration procedure I use, many variations of which can be found online.

Free Fatty Acid Content Analysis

Equipment:

+ Scale (0.1 g accuracy or better)
+ 10 ml self-zeroing burette
+ Stir plate and stir bar
+ 250 ml beaker

Reagents:

+ Oil
+ Isopropyl alcohol
+ Phenolphthalein (0.1%)
+ Potassium hydroxide (0.1 N or 0.01 N solution in deionized water)

First you run a blank titration (no added oil sample) to assess the potential acidity of your isopropyl solvent. I have virtually never had acidic isopropyl, but it can happen, and it will throw off your results.

Blank Titration Procedure:
1. Add 20 ml isopropyl alcohol and 6 drops phenolphthalein indicator to a 250 ml beaker.
2. Add 0.1 N KOH solution by drop until the solution turns pink.
3. Record the amount of KOH required to turn the solution pink. If the solution stays pink after adding only one drop, record the amount as 0. This is the value for **B** in the analysis equation.

Sample Titration Procedure:
1. Place a 250 ml beaker on the balance and tare weight to 0.0 g.
2. Add approximately 8 to 12 g of sample and record the weight to the nearest 0.1 g. This is the value for **W** in the analysis equation.
3. Add 50 mL of isopropyl alcohol, 8 drops of phenolphthalein indicator and a magnetic stirrer to sample beaker.
4. Fill the burette with 0.1 N potassium hydroxide solution—note the starting volume in your burette.
5. Start the stirrer to mix the solution without splashing or swirl to mix.
6. Add KOH solution to beaker by drop until a pink color appears and remains for 30 seconds.
7. Record the amount of potassium hydroxide solution needed for the color change. This is the value for **A** in the analysis equation.

Analysis:
To calculate the percent FFA in oil, use the following equation:

$$((A\text{-}B) \times N \times 28.2) / W = \% \text{ FFA in sample}$$

A=. volume of titrant solution used to produce a color change in the sample;

B= volume of titrant solution used in the blank;

N= normality of titrant (0.1 N in this procedure);

28.2 is the molecular weight of oleic acid, the most common FFA in our typical feedstocks;

W= measured weight of the sample (grams)

Single Stage Base Reaction

Because transesterification is an equilibrium type reaction, both reactants (glycerides and methanol) and products (methyl esters and glycerin) are present at all times. This is why we use more than the minimum methanol required and why doing the reaction in more than one step can help push the reaction towards more complete conversion of the feedstock into biodiesel. That being said, fuel that meets the ASTM specification for biodiesel can routinely be produced using a very simple single stage reaction using a base catalyst.

In a single stage base reaction, we mix our base catalyst, KOH or NaOH, with methanol forming potassium or sodium methoxide and a small amount of water. We'll then add that to our heated and dewatered feedstock in one batch and mix until the reaction has proceeded as far as it will go. We then let the glycerin cocktail settle to the bottom of the reactor, drain it off, test for conversion completeness and then remove the impurities via water washing or "dry washing" techniques.

Catalyst Calculation

For a single stage, I advise using 9.0 grams of KOH per liter of dry feedstock adjusted for KOH purity and FFA content. If your KOH is 90% pure, which is a common form, you'll need to adjust that baseline amount up to 10.0 grams per liter. To this add the FFA percent calculated from your titration.

Here's an example. Our oil titration indicated that our feedstock was 3.0% FFA content. Our KOH is 90% pure. Our batch size is 100 liters (26.4 gallons).

KOH grams per liter of feedstock = 9 grams per liter adjusted for 90% purity is 10 grams per liter (9/0.9 = 10) plus 3 grams per liter added to adjust for FFA content = 13 grams per liter.

So, for our 100-liter batch, we'll add 1,300 (13 grams × 100 liters) of KOH to our methanol to make our methoxide.

This amount of KOH is on the high side of many home-brew recipes—my rationale is that I'd rather produce higher-conversion fuel than get a slightly higher yield. The other factor in homebrewing is the sometimes less than optimum turbulence in mixing, something a little more KOH can help make up for.

Methanol

I advise using 22% methanol by volume of feedstock when preparing your methoxide. Remember, methanol and oil don't mix well. If you have very good mixing, you may be able to cut down on this amount, but this is the place to start. For each liter of dry feedstock, we will use 220 ml of dry methanol. For our 100-liter batch, we'll need 22 liters (5.8 gallons) of methanol.

Mixing the Methoxide

Remember to use your personal protective equipment when mixing your methoxide. I advise a full-face shield with organic vapor canisters, as detailed in the Safety chapter.

When mixing your methoxide, **always add your KOH to your methanol, never add methanol to KOH.** This reaction is exothermic, which means it produces heat. If you make the mistake and start pouring methanol into KOH, you will create a lot of fumes and heat, and it can quickly get much more exciting than you'd like.

The important thing to watch in the methoxide reaction is that the catalyst reacts completely. This is easy to do with a visual inspection. If there are still white flakes in suspension, the reaction is incomplete. Shake it harder. Or pump it through again. Do not use an electric drill with a stirring extension. That's another good way to have your methoxide vessel explode.

A pump, along with a static mixer, is an excellent way to make methoxide. In a sealed metal vessel, add methanol and catalyst, close the lid and send the mixture around and around. Often the shear of the pump will be enough to provide the mixology required to make methoxide.

Reaction Kinetics

This is where the reactants meet your process equipment. You will need to develop your process to optimize your fuel making for your particular setup. Remember the time-temperature-turbulence concept will control the rate of your reaction, so your reaction time will depend on the temperature of your feedstock and the mixing capability of your system.

Typical reaction temperatures are between 131°F and 140°F (55°C to 60°C). The reaction is slightly exothermic, so your temperature will rise slightly as you begin to process. We want to stay under the boiling point of methanol, which is 148.5°F (64.7°C) at sea level. The higher your elevation, the lower the boiling point, so make sure you know what it is in your shop and don't heat your oil over it. I suggest staying at least 10°F

(5.6°C) under the boiling point of methanol in your area as an absolute minimum. This may mean your reaction will take a little longer, but the trade-off in safety is well worth it.

Typical reaction times are 90 minutes, and then mixing and heating can be turned off and settling of the glycerin phase begins. I suggest settling overnight, when possible, in order to get good separation and therefore a better yield. At the end of your mix cycle, perform a 27/3 test (see page 72) to ensure you have obtained good conversion. If it fails, meaning you can see some fallout in the test tube, you'll need to reprocess by running a second stage reaction. More about this in the Troubleshooting chapter.

Two Stage Base Reaction

This reaction will help you push conversion a bit farther with perhaps better yield or the use of less methanol and catalyst. Transesterification is an equilibrium type reaction; this means that its progress is hindered by the presence of products of the reaction. In our case, as triglycerides are reacted into methyl esters and glycerin, the increasing amount of glycerin present inhibits the progress of the reaction towards converting triglycerides into biodiesel.

In a two stage base reaction, the idea is to split your methoxide into two parts, typically using 80% of the volume in the first stage and the remaining 20% volume in the second stage. In between the two stages, you turn off mixing and allow the contents of your reactor to settle for 30 to 90 minutes and then drain off the glycerin bottoms created by the first stage reaction. This glycerin will also include most of the water introduced with your methoxide, as well as the water produced by the saponification of the FFA present in your feedstock. By removing the glycerin and water from the reactor, you are able to push the reaction farther in the second stage than you would have been able to using a single stage base reaction.

After removal of the glycerin bottoms, the remaining 20% of your methoxide is added, and the mixing is restarted. The second stage reaction is mixed for another 60 to 90 minutes and then allowed to settle at least an hour and preferably overnight. At this point, you should perform a 27/3 test to ensure your fuel is well converted.

Acid-base Reaction

I am not a believer in the use of this reaction in a homebrewing environment unless you are using a BioPro™ or similar equipment. Sulfuric acid is nothing to fool around with; it is extremely dangerous to your health.

An acid-base reaction is called for when using feedstock that contains higher levels of FFA (>5% FFA) or to maximize yield on oils with moderate FFA levels (~1% to 4% FFA). The acid in the reaction converts the FFA present directly to methyl esters; no soaps are formed. This is accomplished via the esterification reaction that, as you might recall from the Chemistry chapter, also produces water. This means that the higher the FFA content of your feedstock, the more water that will be produced via the acid reaction and the worse that will be for the subsequent base reaction. This then requires the removal of the polar acid/methanol/water phase between reactions, adding to the complexity of the process.

Sulfuric acid will severely corrode mild steel and black pipe. This is why the BioPro™ uses stainless steel in its construction. Unless you specifically design your system (tank, piping, pumps) to handle sulfuric acid, I would suggest that it's much easier, cheaper and safer to find a supply of lower FFA oil than it is to plan on routinely doing an acid-base reaction process. However, some homebrewers are successfully using an acid-base process, but I certainly wouldn't categorize this as a "beginner" or "simple" strategy.

The BioPro™ processor uses this method very effectively. I highly recommend their equipment and process for backyard enthusiasts, or for any biodiesel brewers that make enough fuel to warrant the expense. I will discuss the BioPro™ more in the chapter on Reactors and Processors.

Backyard Economics

A variety of reasons leads people to want to make their own fuel, and the desire to save some money at the pump certainly ranks among them. How much money you'll save depends on a number of factors. You'll also need to spend some money on equipment, and you'll be investing your time, which also comes at a cost. This chapter will help to shed some light on the costs and benefits of your fuel making. The economics of your particular situation will be unique to you, but they fall into some common categories. We will use one gallon of biodiesel as the cost basis for our analysis.

My advice is to start this exercise by figuring out how much fuel you want to make, based on how many vehicles you'll be fueling, how many miles they'll be driving and their fuel economy (miles per gallon, or mpg). Let's use US averages for this example. According to the US Department of Transportation Federal Highway Administration (FHWA), the average American driver logs 13,476 miles each year. We'll round that off to 14,000 miles for easier math. Let's assume that there are two drivers in our household, and that both vehicles are diesels that get 30 mpg. This means the two drivers will drive a combined total of 28,000 miles in a year using 933 gallons of fuel. At today's (October 2014) price of $3.75/gallon, that's about $3,500 per year in fuel cost.

Before investing a dime in your setup, make sure you can lay your hands on 1,120 gallons of vegetable oil. Recall that feedstock is the great limitation to biodiesel production—and this restriction includes the backyard.

Cost of Materials

The materials you use for each batch of fuel include your collected fryer oil, methanol and your catalyst. Estimating the cost of your feedstock can be difficult, but includes the following:

+ Payment (if any) to the restaurant for the fryer oil
+ Fuel used to drive to collect the oil
+ Energy used to transfer, settle and dewater the oil
+ Fuel or energy used to dispose of schmutz (the food particles and goo you can't use)
+ Value of your time

It's easy to know how much you pay for the oil, if you pay anything. Calculating your fuel used to pick up the oil is really only significant if you are driving out of your way to do so. The amount of energy used is also fairly inconsequential, and so we are left with putting a value on your time, which is subjective by nature. I would argue that if the time spent collecting oil each month comes out of time you'd otherwise be staring at the TV, then using your time to collect fryer oil is actually a benefit to you. If, on the other hand, you would be billing more hours at your law practice, then you could easily value your time at a couple of hundred dollars per hour.

For our exercise, let's assume you do not pay the restaurant for the oil, you pick it up during your normal activities, you are not a lawyer, and you should watch a bit less TV—so your cost of feedstock is essentially free.

The quantity of oil needed will depend on the quality of the oil available. It is best to pick up used oil in the cubies it originally came in. About 1.2 gallons of oil produces 1 gallon of biodiesel due to yield loss in dewatering and transesterification. This means you'll need to collect about 1,120 gallons a year of oil (933 × 1.2), or 93 gallons a month or 22 gallons a week. That's about 4.5 cubies every week or what four restaurants with a fryer will generate on average.

Based on the numbers in the Gathering Ingredients section, methanol bought in a 55-gallon drum will cost about $0.90 per gallon of biodiesel produced. KOH, the recommended catalyst, will cost about $0.33 per gallon of biodiesel.

These material costs total about $1.23/gallon for the biodiesel produced, before you pay any road taxes due. In North Carolina, my home state, the first 2,000 gallons of homemade biodiesel are exempt from state tax, so only the US federal tax of $0.244/gallon applies.

This totals to $1.47/gallon of homemade biodiesel versus $3.75/gallon for petroleum fuel at the pump—a savings of $2.28/gallon! In one year, my wife and I could save around $2,127!

Cost of Your Equipment

To get into fuel production, you'll be investing in equipment for processing, dispensing and lab analysis—and Oil-Dri! Based on my experience, you'll spend between $600 and $11,000 on equipment (Appleseed to BioPro™). Given our annual savings in the example, your payback period is between about 4 months and 5 years, both entirely acceptable, depending on your financial situation.

Feedstock Collection and Pretreatment

Think of your operation as having a beginning, a middle and an end. The beginning is feedstock aggregation. This is best done in a heated tank, or two. You've collected feedstock, and you need to begin by getting the water out of it. Water will naturally fall out of grease over time, but it falls out much more readily with time and heat. Boiling water out of feedstock also works, but it tends to be a losing game from an energy perspective, and it can also be dangerous.

In the early days of Piedmont Biofuels, we had a Grease Warming Zone, a south-facing passive solar room with barrels and totes filled with feedstock. Using the sun is slower than using hot water, or a boiler, but it works.

After a period of settling, the water falls to the bottom, and the chunks tend to float to the top. The ultimate feedstock-settling tank has a port in the middle from which you can draw the dewatered/chunk-free grease.

Feedstock Collection

Grease collection at the backyard scale can be accomplished in a number of ways depending on the desired scale. The average grease collection per restaurant is about 40 gallons a month; some in your area may generate as little as 10 gallons, others will produce significantly more. Contact restaurants in your area to assess your ability to get their oil, and expect many different responses to your query. You will need to find sources totaling about 50 gallons per month per vehicle; this would be two or three restaurants.

The simplest approach to getting good oil is to remember that restaurant managers have lots of things to think about every day, and oil disposal is not usually chief among them. Start by finding out what they currently do with their oil, and if they are willing to give it to you instead. Generally, restaurants just want their oil to go away quickly, cleanly and without drama or spills. If you can help them solve a problem with their current arrangement, all the better. Remember that you are offering a service in exchange for the resource. Their oil will need to be picked up routinely, without fail, rain or shine, summer and winter, and when you are on vacation. Period. Drop the ball once, and you may well lose the account; drop it twice, and you're probably out.

Find out what type of oil restaurants use. Most fryer oil is soy, canola or a blend that may also contain cottonseed or palm. As long as it is low in palm oil, it will be fine for your use. Also make sure that they don't mix in any grill scrapings or soap cleaners; these are messy and don't make for good fuel. Ideally, the restaurant will return the used oil to the cubies that it came in. This helps keep other contaminants from getting mixed in prior to collection. It also is an easy way to transport the oil back to your shop, since each of these will weigh less than 40 pounds. Many smaller restaurants use this method, and it's worth seeking them out.

Most restaurants collect oil in 55-gallon drums, which not only hold a lot but can be accessed outside of business hours. The challenges to the small-scale collector are that the drums are often unsecured, and may contain contaminants, such as rainwater, trash and used motor oil. However, they can be secured with a locking top, which can be purchased online, that prevents most of the trash as well as oil theft.

The other challenge with 55-gallon drums is that they weigh about 400 pounds when full of oil, making them hard to move without a drum dolly and a truck with a lift gate. Another option is to pump or suck the oil out of the drum into a container on your vehicle, like commercial collectors do, with different sorts of positive displacement pumps. This means spending more time behind the restaurant, blocking other deliveries and spilling on occasion. You will also need to supply power for your pump, 12 volt if coming from your vehicle, or access a power plug on the building. Gasoline-powered pumps are not a good choice due to the noise they create, and gas-fired positive displacement pumps are not the norm. If you plan to use drums you need to have a truck with a lift gate, or a truck (or trailer) with a pump and hoses. Oh, and don't forget to pack a spill kit.

Pretreatment Options

After collecting your oil, you'll want to get it reactor ready by settling out any water and schmutz and then drying it more if necessary. If you remove the schmutz and dewater the oil when you first bring it home, this will reduce the rate of oil degradation due to hydrolysis and bacterial action.

Settling the oil is the first step. If you pick up the oil in cubies, it will have already started to settle. Take a look at the

quality of the oil; ideally any water and most of the food particles will have already settled to the bottom. I suggest pouring
this oil into another container; a metal or plastic drum works
fine for this. Pour the oil slowly, taking care not to stir it up,
and don't pour in any water or schmutz visible in the bottom
of the cubie.

Allow the oil to settle at least 24 hours before you load your
reactor by pumping oil from the top of the container with a
"stinger." A stinger is a piece of pipe (PVC is fine) attached to
a hose on the suction side of your transfer pump that allows
you to pump off the top of a barrel, lowering it in the process.
This lets you transfer the cleanest oil and avoid the water and
schmutz that invariably will concentrate at the bottom.

Heating up your oil will greatly decrease the time needed
to settle. One good strategy is to use a heated oil settling drum
and then transfer the preheated and settled oil directly into
your reactor for processing, rather than allowing it to cool and
then reheating it. A basic heated settling drum can be made
from a 55-gallon metal drum equipped with band-type drum
heaters. You can find various designs online. For a particularly
good one, see Utah Biodiesel Supply (utahbiodieselsupply.com
/dryingtank.php).

Filtering used fryer oil is generally not advised if you are
making biodiesel from it. Heated settling works well, and filtering is messy, time-consuming, expensive and unnecessary. Any
remaining particulates that get into your reactor will drop out
with the glycerin bottoms.

Pumps and Hoses

What you want at the beginning of your system is an easy way
to transfer feedstock from your collection vehicle to your initial dewatering tank. This is a good place to talk about various

types of pumps and their uses. In fuel making, we move liquids around a lot, some flammable, some combustible, some viscous (thick) and some not. You will probably end up with more than one pump in your operation, and knowing how to choose the right one helps.

First, make sure that the pump you choose is compatible with the material you intend to transfer. Biodiesel is a great solvent, so any pump needs to be able to withstand it. However, many pumps are used for purposes other than what they were designed for. Best to choose one based on testimonials of people who are using your type of reactor. For example, the Appleseed reactor, which uses a "clear water" pump as its main pump, is designed for water, not biodiesel. It has, however, worked fine when used as described in the Appleseed procedure. Viton is the pump seal of choice for biodiesel use.

Most importantly, make sure that the transfer system you choose for methanol and methoxide has been proven to operate with those flammable and corrosive liquids. Air actuated pumps work great for this; I have used an air-operated double-diaphragm pump with Teflon (PTFE) seals for methanol transfer for years with good results. Avoid using Nitrile, Buna-N and Neoprene with biodiesel. Get a good recommendation within your price range from product literature or from someone who has used that specific pump for the application.

In addition to material compatibility issues, determining the style of pump that meets your need is important. Self-priming or positive displacement pumps can suck up air before they get filled with liquid, like sucking up an empty straw. In another type of pump, a roto-dynamic pump, kinetic energy is added to the fluid by an impeller, and the liquid is flung out of the pump by this force. Centrifugal pumps are a common example of this type. The table below shows some of the benefits and drawbacks of each type of pump, along with some examples of those used in biodiesel production.

Pump Type	Pros	Cons
Positive Displacement • Diaphragm • Gear • Sliding vane	Self-priming Higher pressure Higher head Less shear	More expensive More dangerous Less tolerant of solids
Roto-dynamic • Centrifugal	Less expensive Safer: lower pressure and suction More shear	Not self-priming Lower pressures Lower head

For transferring grease, diaphragm pumps are great since they are much more tolerant of chunks than most positive displacement pumps. They are not as fast as most centrifugal pumps but probably make up for it with fewer clogging episodes.

Roto-dynamic pumps tend to have more shear (think internal mixing) than positive displacement pumps, so they are better suited for a reactor using its pump to mix the reagents. Shear is not your friend if you have too much water in your mix (emulsion!), so impeller or paddle mixing is a superior technology, but also more technically challenging and probably more expensive to do safely and well.

Sliding vane pumps are the choice for commercial fuel distributors since they are self-priming and fast, but they don't stand up well to any bits of solid in the line. So as you can see, pump choice matters, and it needs to be informed by your needs, budget, available time to make the transfer, material compatibility and its role in your system.

Your hose choice is much like your pump choice—it needs to be based on what material you're transferring and your

budget. Hose manufacturers have become much more understanding about their product's compatibility with biodiesel.
My hands-down favorite is Goodyear Flexwing VersaFuel
Flex hose, but it's expensive. A company called GPI is offering
a B100-compatible hose for a much more reasonable price.

It's a good idea to leave your hoses and pumps empty when
not in use; this not only reduces the chance of spills but also
prolongs the life of the equipment. I have also found that keeping the outside of my fuel hoses clean greatly extends their life.
Because many hoses do not use the same material on the inside
as on the outside, your hose may resist biodiesel internally, but
it will get eaten up on the outside. I try to never drag my hoses
through oil or biodiesel, and if I do, I clean them with soap and
water before storing them.

I am a great believer in camlock fittings on the ends of
hoses—one male and one female per hose (seems simple
enough doesn't it?). They connect and disconnect quickly and
securely, and when stored coiled and hooked together, they stay
contaminant free.

A last point on using pumps and hoses—keep cross-
contamination in mind. I reserve one hose for methanol transfer and one for finished fuel transfer that are never used for
anything else. The same with pumps. Be aware that when you
load your reactor with oil, some will remain in the hose and
pump used in the transfer.

You can either use a different hose and pump to transfer
the finished fuel or pump some of the finished fuel through
to clean out the unreacted oil and add that contaminated fuel
back into your next batch of biodiesel, or live with contaminated fuel. Perform a 27/3 test (see page 72) on your fuel
storage tank to assess contamination with triglycerides.

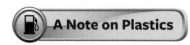

A Note on Plastics

Don't use plastic. Don't buy a system that uses plastic. Plastic does not hold up well over time with heat and generally cannot be repaired when it gives out. There are plastic welders, and there are products for mending plastic—but none of them work very well.

At one point, I thought a 5-gallon water bottle, from atop a watercooler, would be the perfect vessel for mixing methanol and catalyst. We call the outcome of this reaction methoxide.

I added the right amount of catalyst, which is typically granular or in powder form, and the right amount of methanol, and I began shaking it back and forth. The plastic turned brittle as the reaction progressed, and heated up as expected, and the vessel dissolved in my hands, providing me with a nice methoxide shower. Since methanol can travel through the skin, you don't ever want to come in contact with it. If you do, strip and hit the shower fast. Some advise that you take a shot of moonshine to activate the liver before the methanol arrives. I used to suspect this advice was merely coming from those who needed a shot of hard stuff after encountering a methanol event (and we do keep some high-percentage alcoholic beverage on hand at Piedmont for this very purpose); however, many MSDS sheets suggest this as a way to minimize methanol poisoning.

Plastic can also store a charge—especially if you are pumping liquids into it. I hate plastics and suggest they be avoided at all cost. Bob correctly points out that methanol frequently ships in plastic drums and that backyard brewers the world over use plastics with methanol. ⬤

Reactors
and Processors

Now we've gotten to the heart of the matter, the middle part of your system. The reactor you choose will inform much of what you do, including how much time and money you invest in fuel making. Countless designs employ a myriad technologies available to the biodiesel homebrewer. All combine time-temperature-turbulence to get the methoxide intimate with the used fryer oil in order to push the reaction towards completion.

Remember that there is not much flexibility when it comes to the temperature for reaction. It should be as reasonably hot as possible without getting too close to the boiling point of methanol at your altitude. The main variables to consider when choosing a reactor and fuel purification system are:

* How much time are you willing to spend making fuel?
* How much money can you afford to spend making fuel?
* How hands-on do you want to be making fuel?
* How hands-on do you want to be designing and building you fuel-making equipment?

There is no right answer here—only one that is right for you. With so many possibilities demonstrated to work, I've combined them into three basic choices:

+ Design and build your own system from scratch.
+ Build your system yourself, based on a proven design.
+ Buy a system designed and built by someone else.

Designing and building your own system can be very rewarding if you like to tinker. On the online biodiesel forums, countless examples of homemade designs range from elegant and efficient to janky and dangerous. If making this choice, I strongly suggest visiting a local homebrewer first (found on the forum) and spend some time helping in their shop to get a feel for the process. You should have some carpentry/welding/electrical skills, as well as experience working with hazardous materials. This choice is good if you have more time than money and aren't necessarily in a hurry to make fuel.

Building your own system based on a proven design is a great choice, particularly if you are on a tight budget. Although more time-consuming than buying an off-the-shelf system, it will save you a lot of money. You can build an excellent biodiesel production system including oil dewatering and methanol recovery for well under $2,000.

Be aware that chemical agitators can deliver a lot of turbulence with very small propellers and rotations per minute. Not knowing this in the early days, I welded a boat propeller to a stainless shaft attached to a 3,500 rpm motor, a big propeller with a lot of power. You could have water-skied behind the damn thing. I then bolted that to a pair of angle irons I had welded to a 55-gallon drum. When fired up, it sounded like a jet aircraft taking off. And it mixed the reactants beautifully, until the shaft bent, taking the inside of the drum up to cherry red, causing the entire unit to explode.

After that, we invested in a chemical mixer that was appropriately sized for a 55-gallon drum. It was 350 rpm with a propeller a little bigger than a silver dollar.

A huge number of proven biodiesel system designs are available, some free. Not everything that's free is worthless, and not everything with a price is worthwhile. The following options have been used for years and enthusiastically supported on the biodiesel forum by people who have used them.

Appleseed Reactor

The Appleseed reactor was created in California early in this century as an open-source collaborative project headed by Girl Mark (Maria Alovert). Based on no requirement for welding, it is made from off-the-shelf, easily available parts and can be operated safely.

Girl Mark used to sell the plans on the Internet and gave biodiesel workshops countrywide based on the Appleseed design. Sadly, she vanished from the backyard biodiesel community, and in doing so, left with a lot of people's money. Her name, long associated with successful backyard production and fuel quality, is now associated with "rip-off," from the people she stiffed.

The heart of the Appleseed is an electric water heater that is modified to become a reactor with the addition of a pump, some piping and valves. It can make ASTM-quality fuel via either a single stage base reaction or a two-stage base reaction. You can build one for about $300 if you find a good used water heater, or perhaps about $600 starting with a new water heater and adding a few bells and whistles.

An Appleseed reactor, an excellent choice for those with more time (and patience) than money, has a worldwide following of very active users who continue to make improvements. Most people can operate it safely by following the proper protocols and procedure. Thousands of these have been built and operated successfully.

Plans to build an Appleseed are available free online, although it's worthwhile buying the definitive guide to building and operating one, compiled and produced by Rick Boggan (b100supply.com). He also offers extensive free information on Appleseed design and operation on his website (make-bio diesel.org). The fact that this design is open-source, is used by so many and has comprehensive online support makes it a winner.

Graham Laming Eco-System Waterless Biodiesel Processor

I first met Graham Laming online back in late 2005 when he joined the Biodiesel Discussion Forum (biodieseldiscussion .com). He designed and built a processor using a propane gas tank in place of a water heater and christened it the Eco-System. This processor, which uses no water washing and recovers methanol, is a bit more advanced to build and operate than the Appleseed. He has continued his design-build efforts, creating a next-generation processor, the GL Push-Pull Eco-Processor, that incorporates vacuum into the process to aid in drying the feedstock as well as methanol recovery. These are both excellent open-source homebuilt systems with large followings. All the details are on Graham's website (graham-laming.com/bd /main.htm) and at Rick's website (make-biodiesel.org).

Murphy's Machines

In 2007, Paul Oliver started Murphy's Machines to offer professional-quality plans for his design of a biodiesel processor. His MM63 all-steel processor is a 55-gallon drum modified to be completely draining via a cone bottom. These plans require that you or someone fabricate the tank by creating a cone from sheet steel and welding it onto the bottom of the drum. This system, capable of producing batches of up to 50 gallons, can cost from $300 to $600, depending on the options you choose. Although a printable electronic copy of the plans costs $99, Paul provides technical support that's worth at least $100 per hour. He has regularly contributed to the online Biodiesel Discussion forum. Many forum participants have successfully used his plans to fulfill their fuel-making dreams. In addition to biodiesel processors, Paul is the inventor of the Super Sucker oil collection tank that uses vacuum to collect oil quickly and painlessly. His designs can be found at murphysmachines.com.

BioPro™ Processor

Off-the-shelf biodiesel systems offer a turnkey solution to biodiesel processing equipment. You will spend a lot more money for this choice, but will spend a lot less time getting going and may also spend less time making every batch. The only turnkey system I can recommend is Springboard Biodiesel's BioPro™. If I got into making fuel on my property, this would be my choice. Fortunately, I live in the same town as Piedmont Biofuels, and the Tami Tank (one of the locations on the B100 Community Trail) sits about 40 yards from where I park my car in my pole barn.

I can't recommend most off-the-shelf systems because they use plastic tanks for the processor, a big no-no in my book. Plastic tanks are typically rated, at best, to operate at 140°F (60°C). This is too close to reaction temperature for my liking. Additionally, they are much more prone to melting and releasing their contents in a small fire than a steel tank is. Biodiesel processors with plastics tanks can and do make great fuel. These systems typically cost at least $3,000; for much less, you can build an Appleseed that is just as capable of making good fuel, and safer as well.

The BioPro™ is a different story. It is well-designed and con-structed from stainless steel. Making biodiesel in a BioPro™, due to its design and recipe development, is like doing a load of laundry. We bought one of their units for the Central Caro-lina Community College biofuels program, and working with it is a dream. They have developed a two-step acid-base process to increase yield and reduce soaps through the esterification step. You can easily dry your oil in the unit prior to starting the batch. It then washes and dries the fuel in the same ves-sel. After loading the machine, you hit the start button and come back the next day to drain the glycerin. After attaching a pressurized water hose, you hit the wash button and come back the next day to finished fuel. Period. In addition to the acid-base process, the BioPro™ uses a propeller-type agitator to mix, something that achieves excellent results. This facilitates good reaction kinetics and makes the unit great for dewatering your feedstock.

The drawback? Cost. A BioPro™ Model 190 that can make 45 gallons or so of on-spec fuel from 50 gallons of oil costs $9,995, plus shipping. This might not make sense for a person fueling a single vehicle, but for a couple, the payback period can be under five years. A group of friends collecting oil and making fuel together could recoup their investment quickly. The unit can easily produce over 7,000 gallons a year, given feedstock and responsible glycerin disposal, that can save over $15,000 at current fuel prices.

That would fuel fourteen 30 mpg vehicles driving 15,000 miles per year. In my opinion, for small fleet users or fuel co-ops, it's the way to go if you can swing the financing. The other benefit of going with a turnkey is technical support. In this regard, Springboard has been exemplary. When we had a pump fail early on, they sent us repair parts free with no questions asked. You can find more information on the BioPro™ processors on Springboard Biodiesel's website (springboard biodiesel.com).

Cavitational Reactors

When Piedmont was designing and building small-scale biodiesel plants, we would often stray far away from our appropriate technology roots. We developed a cavitational reactor where all the reactants would arrive at a giant pump at exactly the right time, under exactly the right pressure, and it would act like a hammer hitting an anvil. The resultant backsplash would eventually become biodiesel.

A number of cavitational reactors are available that can make good fuel. Although they do work well, they are overkill for the backyard.

Commercial-scale producers have played around with heterogeneous catalysts—embedded in ceramics in order to speed reaction times and reduce costs. Others have embedded microwaves into the reactor. I would ignore most of this technology and focus on a reactor that allows for lots of contact with the reactants.

Continuous flow versus batch processing is a constant conversation in biodiesel, but it is largely moot for backyard and community-scale producers. Since most of these work with a variety of incoming feedstocks, they will always be "batch" in nature.

Purification and Drying Strategies

There are a few different ways to remove the contaminants from your methyl esters after separating them from the glycerin bottoms. They fall into two broad categories—water washing and dry washing.

Purification Strategies

Water washing works well because the contaminants in well-converted fuel are all polar, and so a polar solvent like water will pull them out. The main drawbacks are the need to dry the fuel afterwards, as well as to treat and dispose of the wash water. Using water is also what makes emulsions possible. Here's a few tried and tested ways to water wash your fuel.

5% Prewash

Many fuel makers add water to their reactor *before* draining off the glycerin as a way to reduce the chance of emulsions. After sampling your fuel with the 27/3 test (see page 72) to ensure you have achieved good conversion, add 5% water by volume of the amount of oil you started with. Adding more risks making an emulsion. Continue to mix for another 10 minutes or so, and then allow everything to settle. The water will add to the polar contents and give the soaps something more to dissolve in so that your phases will separate faster and you will have significantly less soap in your biodiesel phase going into your first real wash. This will make washing easier and reduce the chance of making an emulsion. The downside to using the 5% prewash is that, by adding water to your glycerin bottoms, methanol recovery is more difficult, and you'll likely need to have a better distillation process.

Static Mist Washing

This most common way to water wash involves atomizing (misting) water, preferably warm, on the top of your fuel without any agitation. Wash tanks can be made from a variety of different and easily available tanks. A 55-gallon poly drum or a 270-gallon IBC are good choices, as is a purpose-bought cone-bottom poly tank that will make it easier to drain all the water from the bottom. Cone-bottom poly tanks can be found at US Plastics Corporation (usplastics.com). Mist nozzles perfect for the task can be sourced at Utah Biodiesel Supply (utahbiodiesel supply.com).

The total volume of wash water required depends on how much soap is in your fuel, the hardness of your water supply and the temperature of your fuel and wash water. Typically, mist washing uses up to 100% volume of water to the amount of fuel. Performing three smaller washes and draining the soapy water off the bottom is more effective than adding all the water without draining.

Test for clean fuel by first looking at the clarity of the wash water collected after the wash and then putting a small amount of the fuel in a jar and shaking it up. Well-washed fuel will separate pretty quickly and leave clear water behind. Follow with a soap titration before starting to dry your fuel to make sure you've removed soaps effectively.

Bubble Washing

Many home and commercial fuel producers also use bubble washing, which is best used after a preliminary static mist wash, as it is a bit more aggressive and can form emulsions. Bubble washing uses small air bubbles from an aquarium pump and an air stone to help wash the fuel. The air stone is placed at the bottom of the tank in the water phase so that the bubbles formed are made in water, and as the bubble rises into the fuel phase, it is still surrounded by a thin film of water that will "rain" back down through the fuel after it pops on the surface. Fitting your wash tank with both misting and bubbling capability is a good idea.

Dry Washing

Dry washing biodiesel removes contaminants through ion exchange. It is a bit trickier than water washing but solves a few problems—no wash water to dispose of and no drying to perform. Dry washing is done by draining off the glycerin bottoms, then removing the methanol from your biodiesel phase because the methanol acts as a co-solvent that allows soaps to remain dissolved in the methyl esters. The fuel then passes through a tank or column containing material that will pull out soaps and a small amount of glycerin to purify it. The media used in dry washing setups can be purpose-bought material like Amberlite™ or Magnesol, or can be scavenged materials like sawdust and wood chips. Each material requires its own set of parameters in terms of column size, flow rates and the soap, glycerin and methanol content of the fuel. If you are considering dry washing, you've got some more homework to do.

One of the challenges with dry washing is that it's hard to get all the methanol out of the biodiesel phase; therefore, your fuel can have a much lower flash point than it should, making it harder on your motor and potentially dangerous. Another concern is the real risk of spontaneous combustion occurring in your media if you don't keep it filled with fuel to remove the oxygen required for combustion. Additionally, you'll need to responsibly deal with your spent media.

Drying Strategies

After removing impurities by water washing, your fuel will look like pulp-free orange juice. Drying your fuel is the final step before putting it in your tank. There is no need to filter your fuel before use if you have been careful not to cross-contaminate it somewhere in the process.

Drying fuel is very similar to dewatering used fryer oil: you can stir it and heat it, and the water will evaporate over time. You can also bubble dry your fuel in the same tank you washed it in by raising your bubblers from the bottom of the tank so they don't pull any settling water up into the fuel mix and running air through your fuel until it is dry. Dry fuel made from used fryer oil looks a lot like apple juice. It should be crystal clear—no cloudiness; you should easily be able to read newsprint through a pint jar of it.

If you use a heated tank for dewatering your oil, you can make a similar one for dewatering your fuel. A 55-gallon steel drum with a band heater and a circulating pump with a fan-type spray nozzle works great. Plans for such a unit are available online. If you bubble wash your fuel, it really is quite simple to use the same tank for drying by raising the bubblers up a bit.

Methanol Recovery Strategies

Methanol recovery is, along with glycerin disposal, one of the most challenging aspects of making your own biodiesel responsibly. Recovering methanol from both the biodiesel phase and the glycerin bottoms is possible to do on a small scale, as demonstrated by Graham Laming's Eco-Processor and by others. Recovering methanol at a purity of 99.5% or better is significantly tougher, usually requiring a distillation unit capable of reflux. This too can be accomplished on a small scale.

It is easier to find a way to get rid of your glycerin by-product without removing the methanol first. The only two ways I know are to get it to a biodiesel plant, which will probably take it for free, or get it to someone permitted to run an anaerobic digester, like those in most municipal wastewater plants. Neither of these options may be available in your area. If so, please do the right thing and design a methanol recovery strategy into your fuel making.

One reason to recover methanol is that it transforms the cocktail from being a hazardous to non-hazardous material. Theoretically, another is so that it can be reused in future batches, not only to reduce the carbon intensity of the fuel, but also to lower methanol costs. Realistically, this is extremely rare—even in commercial production.

Methanol, being hydrophilic, will grab onto all the water in the system—whether moisture introduced in the feedstocks, chemical water generated during the reaction or ambient moisture from the atmosphere. That means that recovered methanol comes back "wet." Recall that moisture is not your friend when making biodiesel. Some folks are able to add their wet methanol into the anhydrous methanol that they buy—thereby slightly contaminating it somewhat—but the best strategy is to not mess around with wet methanol in the first place.

Bob and I anguished over the topic of methanol recovery for this book. At one point, it was going to be an entire chapter. We both understand methanol recovery; we've both done it, but always on a commercial scale. As this book took shape, we realized that while methanol recovery is feasible in the backyard (Girl Mark successfully achieved it in her Appleseed reactor design that was based on a water heater), very few folks in the backyard bother to mess with it.

Before scrapping the chapter, we talked to some consultants and experts and sent out feelers to the community to see if anyone was successfully doing methanol recovery in the backyard. What we got back was radio silence.

The best idea is to dispose of your glycerin cocktail "methanol in." Once it is out of your way, you can go back to making another batch of fuel to keep you driving down the road.

You can make biodiesel in a 55-gallon drum with a canoe paddle. In the Control Room at Piedmont Biofuels, we have a canoe paddle mounted on the wall—like a gun you might hang over a mantle—to remind us of appropriate technology.

Fuel Quality and Backyard Analytics

Some people in the biodiesel industry love to bash backyard brewers on the basis that they "are not making good fuel." That is sometimes true. And sometimes not. The same can be said for commercial producers, by the way, in an industry with a history of fuel problems.

It is entirely possible to make high-quality fuel in the backyard.

The ASTM specification for biodiesel is D6751. Some of it, like knowing your distillation curve, is nonsense and can be ignored. And some of it is critically important. In order to understand the specification, it's helpful to understand the history of biodiesel.

In many ways, the biodiesel industry took a lesson from the ethanol industry that emerged in the 1970s. Back when the Organization of Petroleum Exporting Countries (OPEC) was holding America up without a gun, the corn lobby stepped in with a new fuel they called "gasohol." We didn't need to worry about far away countries that didn't care for us much; Big Corn had come to the rescue. All you needed was a new pipeline, and a new pump and a new engine, and you were all set to run on ethanol.

By the time biodiesel emerged decades later, created by the farmers and traders of "Big Soy," they elected to blend perfectly into the diesel fuel system. No need to go under the hood. No need to change a thing. Biodiesel was designed to simply "get along" with the petroleum industry. And because of this, there are some weird things in the biodiesel specification, things the petroleum industry requested that the biodiesel industry granted out of a desire to "fit right in." Which means some parts of ASTM D6751 are essential; others are a waste of time.

There are two reasons for you to become proficient with some analytic methods to support your backyard fuel making. First, you want to be satisfied that the fuel you produce meets the specifications to the level you desire. We believe there's absolutely no reason that your homemade fuel shouldn't meet the commercial spec—ASTM D6751.

The second use for analytics is probably even more important to you—to support your process development for producing fuel. It's a lot more satisfying (and easier!) to produce good fuel than it is to muck around with underconverted or wet fuel that's prone to forming emulsions.

We are assuming you simply want to make fuel, fill up and drive. That's much more fun than trying to wrestle a bad batch of fuel into something you can use.

Quantitative versus Qualitative Tests

Let's take a minute for some background information on two types of lab tests. Quantitative tests are those that result in a number or value. Titration is a quantitative test since the result is numerical (quantified). Qualitative tests are those that produce a non-numeric result like "pass" or "fail". Both types of test are valid and useful when characterizing the nature of your fuel.

Key Backyard Specifications

In order to ensure you've made good fuel, you'll want to become proficient with a few fuel quality tests that should be routinely performed. Although commercial biodiesel has a long list of required specifications, only a few are likely to cause troubles if you don't meet them. Some of them affect the way your car will run on homemade fuel, and others will affect the shelf life of your product and ultimately degrade your fuel to a point that you wouldn't want to use it. Fortunately, we really only need to address these four fuel parameters:

+ Conversion and removal of glycerides
+ Moisture (water) content
+ Soap content
+ Cloud point

Conversion

Getting good conversion of your feedstock into biodiesel is the single most important factor in your fuel making. If you follow the recipes, and have a reactor with decent mixing, there's absolutely no reason that you should not be able to make "on-spec" fuel right out of the gate.

Poor conversion means that the triglycerides in your feedstock are not converted into methyl esters, remaining as mono-, di- or triglycerides. These unconverted glycerides can wreak havoc with your fuel-making endeavors in a number of ways.

Mono- and diglycerides are emulsifiers. Any in your fuel will make it more likely to create an emulsion when washing, resulting in more of your fuel being discarded with the wash water. Their presence will also make it harder to get your fuel dry, and any moisture that enters your storage tank will be harder to remove. This tends to make your stored fuel unstable due to the increased possibility of hydrolysis.

Unconverted triglycerides will make your fuel more viscous (thicker), resulting in poorer cold flow performance as well as more strain on your fuel injection pump.

The bottom line is that you **can** run underconverted fuel in your diesel motor, but our advice is not to. It's just too easy to make well-converted fuel, and the many benefits are worth the little extra effort.

Measuring conversion is easily done with some methanol, which you already have on hand, and some of your fuel. The 27/3 test is a qualitative test that measures the conversion quality of your fuel by assessing its ability to completely dissolve in methanol. Since triglycerides are not soluble in methanol, any unconverted triglycerides will be visible in your test vial. If you can see any amount of "drop out," your fuel is not completely converted and should be reprocessed by adding a bit more methoxide and reacting it a bit longer.

 **27/3 Test**

To perform a 27/3 test, put 27 milliliters of good-quality methanol in a test tube or centrifuge tube. Graduated centrifuge tubes work well, can be washed and reused many times, and can be purchased online. Use methanol that is at room temperature, about 70°F (21°C). Add 3 ml of fuel; washed or unwashed will work. Invert the tube once, then turn it right side up again and look to see if any of the biodiesel is not dissolved and is falling towards the bottom of the tube. If any fuel didn't completely dissolve in the methanol, it is underconverted. Well-converted fuel will completely dissolve in methanol. ⬤

There is a correlation between how much drop out you see versus how underconverted your fuel is—the more precipitate, the less conversion you have achieved. Some people have gone to great lengths to quantify the amount of drop out in order to come up with a "percent conversion" number. I have used the 27/3 hundreds of times and have compared results to those given by a gas chromatograph (GC) used by labs to measure conversion very accurately. In my experience, fuel that showed no drop out passed the ASTM specification for free and total glycerin by GC, whereas samples that failed the 27/3 also failed on the GC.

The best way to make fuel is to develop your process (ingredients and equipment capability) to make on-spec fuel every time, rather than trying to come up with some type of percent conversion by quantifying 27/3 results. Getting good conversion is as simple as using dry feedstock, dry methanol and dry catalyst, and getting enough heat, mixing and time to get the job done.

Moisture Content

Water in fuel is never a good thing. In diesel motors, it can lead to fuel degradation and injector damage. Many diesel vehicles have a water separator as an integral part of their fuel filter. This way, water is captured and drained off prior to getting into the motor itself. Petroleum diesel fuel typically holds 200 parts per million (ppm) or less of water in solution. Warm fuel can hold more dissolved water than can cold fuel. Biodiesel can hold up to 1,500 ppm water in solution; therefore, we need to be careful about how we dry our fuel after producing it, as well as how we store our fuel in order to keep it dry. "Dry" fuel should contain 500 ppm (0.05%) water or less.

We can measure the water content in biodiesel in two relatively cheap and easy ways. The first test should be a visual check of the fuel in a clear glass jar. It should appear "clear and bright," and look a lot like apple juice, if made from waste cooking oil. There should be no visible haze or sediment, and when holding the jar of finished fuel up to some newsprint, you should be able to see the print easily and clearly. If there is any visible haze or sediment, you need either to continue to dry and/or to filter your finished fuel.

Another method of assessing water content is to invest in a test kit that uses calcium hydride to measure the moisture in your fuel. One popular model, called the Sandy Brae test, can be purchased online for $225 to $350, depending on the model.

Soap Content

Soap, a by-product of your fuel making, is created by the saponification reaction that occurs when your catalyst comes in contact with FFA in your feedstock. The lower the FFA content in your feedstock, the less soap will be created in your fuel.

Soaps are emulsifiers, so they will make it harder for you to dry your fuel after water washing, and can cause emulsion problems in your fuel tank in the presence of water.

The ASTM spec for biodiesel does not have a soap limit per se, but since a soap molecule is a combination of an FFA and the metal (sodium or potassium) in your catalyst, high soap levels in your fuel will cause it to be out of spec on sodium or potassium content. Commercial producers try to keep soap levels under about 50 ppm, which should also be your goal.

Test for soap using the following titration procedure. Although it's similar to titrating for FFA, we use a hydrochloric acid (HCl) solution, rather than a basic (KOH) one, and add bromophenol blue as a color change indicator, as it works down in the acidic range of the pH scale.

First we titrate the solution to account for the presence of residual catalyst, using phenolphthalein as the color change indicator. In washed fuel, the level of residual catalyst should be zero. We then continue to titrate using HCl, which breaks the soap molecules into FFA and a salt, until the bromophenol blue color change indicates all soaps have been accounted for. This method lets us accurately assess the level of soaps in our finished fuel.

Equipment:
- Scale (0.1 g accuracy or better)
- 10 ml self-zeroing burette
- Stir plate and stir bar
- 250 ml beaker

Reagents:
- Biodiesel sample
- Isopropyl alcohol
- Phenolphthalein (0.1%)
- Bromophenol blue (0.04%)
- Hydrochloric acid (0.1 N or 0.01 N)

Procedure:
1. Measure about 10 g of unwashed biodiesel into a 250 ml beaker and record the weight to the nearest 0.1 gram. (Use about a 20 g sample for washed biodiesel.) This is the value for **W** in the analysis equation.
2. Use 0.1 N HCl in your burette. (Use 0.01 N HCL for washed biodiesel, as this will allow you to measure low soap numbers more accurately.)
3. Add 75 ml of isopropyl alcohol to the beaker. Place the beaker on the stir plate and stir without splashing.

4. Add 5 drops of phenolphthalein to the alcohol/biodiesel solution. If the solution turns pink, add HCl until the solution loses the color. Record the volume of titrant used to clear the pink color. This is the value for **A** in the analysis equation. If you are measuring washed fuel, you will likely not see a pink color when adding the phenolphthalein and therefore won't need to add any HCl to clear it. In this case, the value of **A** is zero.

5. Add about 15 or 20 drops of bromophenol blue to alcohol/biodiesel solution—you want an easily detectable blue/green color.

6. Add HCl titrant to solution until it turns from blue/green to bright yellow and remains yellow for 30 seconds. Record the volume of titrant used to effect the color change. This is the value for **B** in the analysis equation.

Analysis:

To calculate the amount of soap in your biodiesel sample, use the following equation:

$$((B-A) \times 0.1 \times 320.56)) / (1,000 \times W) = \text{grams of soap in your sample}$$

A= volume of titrant solution used to remove pink color in first titration (usually 0);

B= volume of titrant solution used to effect the color change from blue to yellow;

N= normality of titrant (0.1 N or 0.01 N depending on which you used. The more dilute the titrant, the more you'll use, but the more accurate your reading will be. I usually like to dial my titration method in to use 0.1 ml to 10 ml of titrant.);

320.56 is the molecular weight of potassium oleate, the most common soap created when using KOH—if you're using NaOH as your catalyst, use 304.45 instead;

W = measured weight of the sample (grams).

To convert your results to parts per million (ppm), multiply final result by 1,000,000.

For example, if you have a 20-gram sample of washed biodiesel made with KOH and use a 0.01 N HCl titrant, you'll only use about 0.25 ml of titrant if your soap content is 40 ppm.

$$(0.25 \times 0.01 \times 320.56) / 20 \times 1,000 = 0.00004$$
$$0.00004 \times 1,000,000 = 40 \text{ ppm}$$

Anywhere under 60 ppm of soap should be just fine for your backyard fuel production. If you are having trouble hitting these numbers, you'll need to refine your washing process.

Cloud Point

A fuel's cloud point is the temperature at which crystallization begins. At this point, your fuel risks clogging the fuel filter and stopping your vehicle in its tracks. This happens as the more saturated methyl esters, as well as longer chain methyl esters, begin to solidify based on temperature. When enough of these "flakes" collect on the fuel filter, the engine won't get the fuel it needs to properly accelerate or climb the next hill. This is known as "fuel starvation," and under the right conditions, it happens pretty fast, leaving you on the side of the road.

There is no ASTM specification for cloud point, other than the requirement to report it to the buyer of the fuel. This way, the user can be informed of the temperature at which the fuel will start to gel and can treat it accordingly. Fortunately, most biodiesel is easily treated to enhance its cold flow properties—simply blend it with regular petroleum diesel.

Your fuel's properties are contingent on what type of oil was used as feedstock. Typical fryer oil is soy or canola based, sometimes mixed with cottonseed, peanut or palm for economic or performance reasons. Canola is the highest in unsaturated fatty acids, while palm is the most saturated. Most used fryer oil will produce biodiesel whose cloud point is around 32°F (0°C). If poultry fat or beef tallow is a feedstock, your fuel will be very saturated, and winter use in climates under 65°F (18°C) is not advised. My advice is to know the approximate cloud point of your fuel, watch the weather forecast and start to blend petroleum diesel into the car's fuel tank when the expected temperature is within 10°F (5.5°C) of your cloud point.

Measuring cloud point is as easy as refrigerating a glass jar of your fuel and checking on it every few minutes for a detectable haze or cloudiness, while you measure the temperature. My infrared (IR) temperature gun is invaluable for this job in fuel making. These guns don't always work well on clear glass or shiny surfaces like stainless steel, so make a spot with a black marker on your vessel to aim it at.

If the fuel is refrigerated long enough to reach the same temperature as the rest of your fridge without cloudiness, move it to your freezer and continue cooling it until you see haze appear—this will be your cloud point. Another method many homebrewers (and commercial producers) use is to keep a glass jar of B100 next to their fuel storage tank or where they park their car, allowing real-time monitoring of the fuel. Remember to update your container with your latest batch periodically.

Here in North Carolina, winter temperatures rarely go below 20°F (−6.7°C), and a B80 mix (80% biodiesel) usually does just fine. Your temperature (and mileage) may vary.

Lab Setup

Investing some time and money to set up lab capabilities is an important part of your fuel-making endeavor. It can be done quickly and inexpensively. Here's some advice from someone who's done just that and a list of some equipment I've used with good results.

+ Lab scale: My Weigh Triton T2 (200 g capacity × 0.01 g), available online at Amazon for $18.
+ Self-zeroing burette: Thomas Scientific part number 2002A01, available online for under $70, thomassci.com.
+ 250 ml beaker: available online from Amazon for under $10.
+ Magnetic stir plates: can be expensive. Utah Biodiesel Supply (utahbiodieselsupply.com) has a nice battery-powered one that comes with a flask and stir bars for under $50. Alternatively, you can often find stir plates on eBay for under $100.
+ Centrifuge tubes for 27/3 test: I use Corning part number 430055, a pack of 50 centrifuge tubes that are 15 ml reusable plastic and costs under $30, available online. Rinse out with soap and water, and you can use them multiple times. I do a 9 ml to 1 ml 27/3 test, so a 15 ml tube works great if you follow that method.
+ Infrared (IR) temperature guns are handy to have around for monitoring reaction temperature and cloud point testing. I like the PE2 model available from TempGun.com (tempgun.com) for $40.

All of this lab equipment will cost about $200, a small investment to be able to ensure producing quality fuel.

In addition to equipment, you'll need to get lab reagents (chemicals). You can make your own 0.1 N KOH solution by dissolving 5.6 grams of KOH in 1 liter of distilled water. Measure both the KOH and the water very precisely to make an accurate solution. A 1,000 ml volumetric flask (along with your scale) will do the trick, available online for about $15.

That's the backyard take.

Commercial producers are required to measure and report impurities in their fuel. Methanol, catalyst, glycerin and unreacted or partially reacted oils can all lead to difficulties with your engine and need to meet ASTM D6751 in order to be legal for on-road sale.

Determining the temperature at which it "flashes" typically involves measuring the quantity of methanol in fuel. A sample of biodiesel is heated up in a device called a flash point tester; the fuel should not flash below 93°C. If it flashes below that, wash your fuel again. Flash point testers range from $700 to thousands of dollars, which is why backyard brewers seldom bother with them.

The amount of free glycerin allowable in on-spec fuel is .02% of the overall mass. And the amount of total glycerin, free plus bound (unreacted oils), is .24% of mass. Measuring this number is typically done with a gas chromatograph, which will report on the glycerol, monoglycerides, diglycerides and triglycerides in your fuel. A used gas chromatograph can cost $5,000 and will need a monthly maintenance budget in order to stay happy.

Commercial producers are required to report their acid number. This can be done through manual titration, which reports results visually, or it can be done with an auto-titrator, which measures changes in the conductivity of the fuel. An auto-titrator can cost $600.

A water and sediment number, another important part of the specification, is determined in commercial biodiesel laboratories with a desktop centrifuge. These range from $300 to thousands of dollars.

Measuring the metal content of fuel is also required. Potassium, sodium and phosphorus (which is not actually a metal) are measured using an inductively coupled plasma (ICP) machine. Even used ICP machines can cost close to $20,000.

You also don't want sulfur in your fuel. This is measured by optical emission spectroscopy (OES) that is typically coupled as an ICP/OES machine. While biodiesel is famous for its absence of sulfur, some feedstocks, such as trap grease, can be extremely high in sulfur and can lead biodiesel to be out of specification.

For $200, and some care, you can make high-quality, on-specification fuel in the backyard. But before you dream of turning commercial, be aware that a single commercial analytical machine can cost more than your entire backyard operation.

Note that not every commercial producer possesses all this gear. Piedmont generates a certificate of analysis for every drop of fuel that ships from its plant. Part of that analysis is generated in our onsite laboratory, which possesses analytical gear; part is generated at Central Carolina Community College (CCCC), which has an extremely well-appointed laboratory; and some is generated by the North Carolina Department of Agriculture, whose fuels laboratory has everything that we and CCCC lack.

Backyarders and small-scale producers frequently send samples to Piedmont for analysis—in order to help guide them in building a process that creates good fuel.

Troubleshooting

Things are always going great, until they're not. Biodiesel is easy to make, and it's also easy to mess up. Fortunately, most biodiesel problems are surmountable, and the feedstock is almost never "ruined." In this chapter, you'll learn how to best avoid some common fuel-making mistakes. Safety issues will be covered in the next chapter—they are also fairly easy to avoid but often not so easy to undo.

Emulsions

Emulsions occur when a non-polar substance like biodiesel and a polar substance like water are able to mix together into a mayonnaise-like glop due to the presence of an emulsifier like soap. This typically happens due to over-aggressive water washing in the presence of underconverted fuel and/or high soap content. So the best ways to avoid making emulsions is to make sure your batch is well-converted by cleanly passing a 27/3 test (see page 72) before you start water washing, and make sure your first water wash is not overly aggressive. A top-down gentle mist for the first wash will do the trick, as will using the 5% prewash technique described in the Purification and Drying Strategies chapter.

The good news is that breaking emulsion is always possible, given time, heat and the right technique. There are a few tried-and-true emulsion-breaking techniques so you can experiment, when the times comes, and determine which works best for you.

Acid Treatment

Acid treatment has been my emulsion-breaking tool of choice. I have used small amounts of phosphoric acid with instant success, both in the commercial plant I managed in Berthoud Colorado and in the lab at Central Carolina Community College. The acid breaks the soap into an FFA and a salt, and the emulsion immediately breaks. I then reprocess (if necessary) and rewash to remove the residual acid, free fatty acids and salt. Vinegar (weak acetic acid) can also be used to break emulsions. Emulsions will be easier to break if the mix is hot, about 100°F (37.8°C). Mix the acid in gently by stirring if possible or by turning on your mixer for a short burst or two. Add more weak acid until the emulsion breaks, then rewash and dry your fuel as normal.

Note: NEVER use nitric acid anywhere in your fuel making. Using nitric acid in a mix containing glycerin can result in an explosive mixture.

Salt Water

Salt water can also be used to break an emulsion. Dissolve regular salt (sodium chloride) like rock salt or table salt in hot water. A cup of salt mixed in a quart of hot water should break up to 50 gallons of emulsified fuel. The premise here is that the salt water greatly increases the polarity of the mix and pulls the rest of the polar material with it, leaving the relatively non-polar biodiesel behind. Add salt water and gently stir until the emulsion breaks, then rewash and dry your fuel.

Glycerin Bottoms

Adding leftover glycerin from a previous batch breaks emulsions as well, like salt water, pulling the water out of the emulsion. You may need to add 30% to 40% bottoms by volume of your emulsions. Again, mix gently while adding glycerin until the emulsion breaks, then allow to settle and decant the glycerin layer. Rewash and dry the fuel.

Poor Conversion

Poor conversion is a problem caused by a variety of factors: typically reacting with wet feedstock, poor mixing, not a long enough reaction time or poor-quality catalyst or methanol.

Wet feedstock, the usual cause of poor conversion, can be overcome by using a Sandy Brae device to test your feedstock for dryness prior to reacting it. The water present should be under 0.05% (500 ppm) or less if possible—the dryer the better.

Poor mixing can be overcome by either running your processor longer in the reaction phase or by increasing the mixing turbulence by adding eductors or static mixers into your equipment setup design if you are pump mixing. If you have successfully made fuel that passes the 27/3 test with your current setup, the problem probably lies elsewhere—check feedstock for water and reagents (methanol and caustic) for water contamination.

Methanol quality can be checked with a hygrometer (see page 97). Caustic is rarely the problem if it has been stored properly—kept dry and tightly sealed. If the caustic is not clumped up or gooey and has been stored well, it's likely fine.

When you end up with an underconverted batch, don't start washing it. Drain the glycerin bottoms and reprocess the batch by adding a bit more methanol and catalyst. I suggest 5% methanol by volume mixed with 1 gram per liter KOH for your methoxide mix. Add this to your batch and reprocess as you would for a single stage. You may not see any glycerin fall out when you settle, but you should have achieved spec level conversion.

Wet Feedstock or Fuel

Keeping your feedstock and fuel dry is fairly straightforward. Put it in the container dry, don't let rainwater collect on the top where it may leak in, and if necessary, use a dessicant breather on your fuel storage tank. I have never had to use this type of equipment because I keep my fuel storage tank tightly sealed when not dispensing and don't routinely store large quantities of. If your tank is more than 55 gallons, and you live in a very humid climate and allow your fuel storage tank to get warm during the day and cool off at night, you may want to monitor the dryness with a Sandy Brae tester and add desiccant to your setup.

Fuel and Feedstock Stability

To keep your feedstock and fuel from going bad due to oxidation, store it properly—no water and temperature changes minimized—and use it 30 to 45 days after production. You can add an oxidation stabilizer to your fuel to prolong its shelf life. I don't recommend additives, as they are expensive, have large minimum order quantities and are entirely unnecessary if you plan your production for use while the fuel is fresh and store it properly. The same goes for feedstock: don't collect more than you can use in 30 to 45 days and store it properly. If you do plan on storing feedstock, settling and dewatering first will help prolong its storage life.

Safety

Knowledge facilitates safety.

The more you know about what you're doing, the easier it is to anticipate and avoid accidents. Although making biodiesel is relatively easy, the chemicals we work with can be extremely dangerous if mishandled. Methanol is highly flammable and toxic. NaOH and KOH are highly caustic and can combust when wet. Even used fryer oil and biodiesel can cause dangerous conditions if handled improperly. Take the time to learn the safe way to handle and store all your materials, and you'll sleep better at night.

Engineered versus Administrative Controls

In order to reduce the hazard presented by materials and processes, controls are used. Engineered controls are those that are designed and built to isolate people from hazards. A lock on your workshop to keep children away from hazardous materials is an example of an engineered control, as is a timer or float switch that ensures that a tank won't overfill and spill.

Although engineered controls require investment, they are usually well worth the expense. And they are not foolproof.

Administrative controls are rules or procedures that help isolate people from hazards. No smoking in the shop would be an example of an easy administrative control to reduce the risk of fire. Properly labeling materials is another good example of an easy administrative control designed to enhance safety. One long-standing rule at Piedmont prohibits beverages in laboratory spaces. If you are not drinking in the space, you are less likely to take a swig of isopropyl alcohol, thereby saving a trip to the emergency room.

A combination of both engineered and administrative controls is typically used to help minimize the risk of personal injury or property damage.

Material Safety Data Sheets

Material Safety Data Sheets (MSDS) exist for all the hazardous materials used in fuel making. These detail all the health, safety and environmental information for the material in question. Search for the material name followed by MSDS (e.g., Methanol MSDS) to find them free online. Take a few minutes and read the details of the hazards, suggested personal protective equipment and first aid measures for each chemical you plan to work with.

Personal Protective Equipment

Personal protective equipment (PPE) like gloves, safety glasses and respirators are useful in guarding yourself from hazards while making fuel. A box of 100 nitrile disposable lab gloves will last for months and cost as little as $8 purchased online. Gloves are a good idea when handling caustic and methanol, and optional when handling used fryer oil and biodiesel. Reusable washable gloves can be great for handling waste oil containers and general work around the shop.

Wear clothing appropriate for your activity: closed-toed shoes, long sleeves and long pants, headwear that will provide some protection from a spill or splash. Consider using a chemical-resistant apron for mixing and transferring methoxide. Avoid clothing or jewelry that is too loose fitting or dangles and creates an opportunity to get hung up on equipment or becomes a distraction at a crucial moment.

Safety glasses or a full-face shield are a must when working with methanol or caustic in any quantity, whether performing a 27/3 test, a titration or mixing methoxide. Get a good pair of safety glasses, goggles or a face shield, depending on what is comfortable. Make sure they fit well and do not need constant attention to keep them from slipping. Regular corrective lenses or sunglasses really don't provide enough protection against splashes—don't count on them as PPE.

Respirators are good to use when pouring production quantities of caustic, as well as when pouring methanol and mixing the methoxide. Caustics can emit dust into the air that can get in your eyes and settle on sweaty skin, causing a chemical burn. A full-face shield with respirators will keep this out of your lungs and eyes and reduce the risk of methanol inhalation. Be aware that due to the small size of methanol molecules, they are not entirely filtered out by a respirator, so design your process to limit the amount of methanol released into the workspace. Change cartridges at least once a year or when they are compromised by exposure to a large chemical release.

Gloves, safety glasses and a full-face shield respirator with organic vapor cartridges can be bought online for about $160, a small price to pay to preserve your health and safety.

Additionally, it's a good idea to get maximum ventilation through your workspace while methanol vapors are present. Please remember that methanol vapors are flammable and toxic, so using an electric fan to try to remove high concentrations of methanol is a bad idea. Much better to not have the methanol escape your process to begin with. If using a fan, place it well away from the source of methanol fumes and blow air into the room rather than using it as an exhaust fan and risk pulling flammable vapors into the fan.

Here are some details on recommended PPE, based on my personal use. These items are available from a variety of online sellers:

+ **Lab gloves:** Kimberly-Clark, Blue Nitrile Powder-free Exam Gloves, Model 53103, box of 100, $24 from Amazon.

+ **Chemical handling gloves:** Wells Lamont 167L Heavy-weight PVC Fully Coated Gloves, $8 from Amazon. These are great for measuring bulk KOH and methanol, as well as methoxide mixing.

+ **Safety glasses:** for everything except mixing and adding methoxide, Uvex S1600 Bandit Safety Glass, black frame, clear lens, $9 from Amazon. These fit over my small prescription glasses.

+ **Full-face mask with organic vapor cartridges:** 3M Full-face piece Reusable Respirator 6900—get the right size! About $100 from Amazon; 3M Multi-Gas/Vapor Cartridge/Filter 60926, 2 pack, $24 from Amazon. Remember, no respirator cartridge will remove methanol vapors entirely. I use this setup in a well-ventilated shop and keep methanol fumes at a minimum.

+ **Chemical-resistant apron:** Steelman 77050 black chemical-resistant apron, $23 from Amazon.

Location, Location, Location

Do not make biodiesel in any building that is attached or part of your house. Period.

If you doubt the wisdom of this advice, please take a moment and Google the words "biodiesel home fire images" and take a moment to reflect. It's best to make fuel in a shop space that is not physically connected to your house. If you plan on making fuel through the winter, you'll probably want a heated space, depending on your climate. When considering heat sources for your biodiesel shop, remember that methanol emits flammable vapors. You can, of course, just turn off your building's heat when you are reacting or washing and may have methanol fuel present.

The folks at Springboard Biodiesel, producers of the Bio-Pro™ processor, have done extensive testing and analysis of its unit's methanol emissions and claim that it can safely be used in a typical home garage. I certainly believe this is possible, having run one of their BioPro 190s numerous times in two different lab spaces at our local community college. That being said, if I had one at home, I would put it out in my pole barn just to be on the safe side.

Oil collection, dewatering and biodiesel production are typically messy operations. Fuel production also drives the need to store caustics and methanol onsite. These are not the kind of activities or materials that fit well in a suburban attached home garage. Your spouse, your kids and your pets will appreciate not having to trip over or come in contact with this when getting out of the car and heading into the house.

Basic Housekeeping

A clean and uncluttered workspace goes a long way to providing a safe environment for your fuel making. Keeping cords and hoses from becoming trip hazards is part of a good design. Store any unused extension cords coiled and hung out of the way. Keep unused hoses coiled and, if possible, hooked end to end to prevent any liquid from leaking out.

It is good practice to entirely empty any hose before storing it to prevent leaks as well as to prolong its life. I like to use camlock fittings since they provide a fluid-tight connection to my processing equipment and allow me to store my hoses with the male end plugged into the female end to prevent leaks and contamination from dust, bugs, etc.

Maintain a clean toolbox by always degreasing your tools before putting them back in the toolbox. They are going to get greasy, but there's absolutely no reason to contaminate your toolbox so that every tool is greasy all the time. It really only just takes seconds to clean tools before replacing them.

Don't let glycerin by-product pile up in or outside your shop. This happens to most fuel makers at some point and has caused a few fines, fires and divorces. It is imperative to find a method of disposing of or using your glycerin by-product before you start to generate it. If at some point your situation changes and you start to stockpile homeless glycerin, I advise you to stop making fuel until you find an alternative.

Keeping your mind uncluttered is important as well. Don't bring your cell phone into your workspace, and if you do, don't answer it. This distraction may be all it takes to help you forget that you have a pump running, causing an overflow spill. Also, of course, drinking, drugs and hazardous materials don't mix.

Spill Prevention and Control

As any of us with grease collecting or biodiesel making experience knows, spills happen fast and clean up slow. It's best to develop habits that help you prevent spills.

Make a habit of reviewing your setup before starting the pump on a liquid transfer, whether it's used cooking oil, methanol, methoxide, glycerin, wash water or finished fuel. My method is to start from the end that I'm pumping from and point with my finger all the way along the route of transfer, looking for loose connections and making sure that valves are in the desired position. I also ensure that the tank I'm pumping into has enough space (headspace) to prevent overfills from the amount of liquid being transferred. I go through these rituals every time before transferring any liquid anywhere.

Make it a habit to not walk away from pumps when they are transferring liquid.

I'm a great believer in Oil-Dri Premium Absorbent ($10 at Lowe's for a 25-pound bag). In my shop, I like to keep a spill kit handy composed of a 5-gallon bucket mostly filled with Oil-Dri, and a dustpan and whisk broom sitting on top. When the inevitable spill happens, I can immediately pour some Oil-Dri on the spill before it gets tracked around the shop. Pour on a generous amount, fully covering the spill and give it time to absorb. Then sweep up the material into the dustpan and reuse it by putting it back into the 5-gallon bucket, unless it is totally saturated. I can usually reuse the material a few times before bagging it for disposal and refilling the bucket with fresh absorbent.

Material Storage

Proper storage of your chemicals is important to promote safety and to maintain their quality. Best storage practices include proper labeling and containers that keep the material safely contained, protect the integrity of the material and are ergonomically sound. Buying chemicals in bulk will save money, but allowing the material to degrade due to bad storage will likely cost you even more. Understanding the nature of the material to be stored, as well as what container materials are compatible, will help you make informed choices.

The materials you will routinely store include used fryer oil, methanol, caustic (KOH or NaOH), waste glycerin, used wash water and finished biodiesel. Let's look at the nature of the hazards and proper storage techniques for each.

Used Fryer Oil

The hazards associated with used fryer oil are its combustibility, the slip hazard when it spills and your risk of injury while moving a heavy load. A 55-gallon drum of oil weighs about 400 pounds—not something to be trifled with. Although not flammable per se, used oil is combustible and would stoke any fire that breaks out in your shop if its container melts and leaks.

Collected used fryer oil should be kept in containers, typically Type 2 plastic (HDPE), like the cubies for transporting new fryer oil a, 5-gallon buckets, 55-gallon drums and 270-gallon IBC's (caged poly totes). Used fryer oil should be properly labeled and dewatered upon arrival at your shop to keep it from further degrading due to biological breakdown of the schmutz (food particles and water) it contains. Once settled and dried, keep your incoming feedstock out of the sun and at room temperature or lower for the best shelf life.

Methanol

Methanol, probably the most dangerous material you'll routinely use, is highly flammable and toxic. It can also be safely stored in HDPE plastic containers or steel drums. Five-gallon plastic carboys are great to transport and store methanol in before use, available online at Amazon or Uline for about $20.

Methanol is hygroscopic, that is, it will draw water to itself. So keep your stored methanol tightly sealed at all times that you are not dispensing it. Otherwise, it will start to pull moisture out of the atmosphere and will become contaminated with water over time. You can test the purity of methanol using a hydrometer, a tool for measuring specific gravity (SG). The SG of pure water is 1.000 at 20°C (68°F), while the SG of methanol at the same temperature is 0.7913. By measuring the SG of your methanol, you can determine the purity with a little math. Here's the formula for use at 20°C (68°F):

$$\% \text{ Purity} = 100 - (B - 0.7913) / 0.002087$$
$$B = \text{SG of your sample at 20°C (68°F)}$$

You can buy a complete methanol test kit online from Utah Biodiesel Supply for $85.

If using a metal 55-gallon drum to store your methanol, you should electrically ground it. Static electricity can build up in your container due to fluid movement, and the resulting spark can cause an explosion or fire. Ground your metal drum by using a clip to the drum that's wired to a properly deployed copper ground rod. If using a pump to transfer your methanol, make sure it is explosion-proof and rated for methanol. Better yet, use an air-powered pump or a hand-crank pump to dispense your methanol.

Caustic (KOH or NaOH)

KOH and NaOH, both strong caustics, are therefore very corrosive. The dust generated when transferring these materials can cause serious burns to your eyes and skin. Once mixed with methanol, the resulting methoxide combines the hazards of both the caustic and the methanol: it is flammable, toxic and a strong corrosive that will cause a nasty chemical burn on contact. Don't treat these materials casually! Always take the time to put on the proper PPE (gloves, apron, safety glasses or face mask), and focus your mind and hands on the tasks at hand.

KOH and NaOH are hygroscopic; they'll readily pull moisture from the air that will ruin their ability to act as a catalyst for transesterification. You need to keep them dry. I've found it's best to store unused catalyst in plastic 5-gallon drums with airtight Gamma Seal lids, found online at US Plastics for about $8, usplastic.com. US Plastics also has some very nice premium buckets in various colors for $8 each.

NFPA Placards

Proper labeling of your stored material may not seem important, but it is. In the event of an emergency in your shop, you'll want the first responders (fire, ambulance, EMS personnel) to quickly understand the nature of the hazards within. Labeling is easy, and it raises your game to a more professional (and safe) level.

The US National Fire Protection Association (NFPA) created a labeling system for hazardous materials that you should use, which you probably already recognize. These "diamond on point" symbols use colors and codes to let first responders quickly assess the hazards present in order to use appropriate measures without unduly risking their own safety or that of others.

Some years ago, I created the following set of NFPA signs for the commercial biodiesel plant I was managing in Berthoud, Colorado.

The colors denote the particular type of hazard. Red (top diamond, gray) indicates flammability, blue (left diamond, black) indicates health effects, yellow (right diamond, light gray)

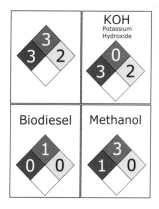

indicates instability, and white (bottom diamond) is reserved for special hazards like oxidizers and water-reactive materials. The number indicates the severity of the hazard: 4 is the most severe, while 0 is no hazard. You can see that biodiesel has a 1 in the red square, as it is combustible but not flammable by NFPA definition. Biodiesel has no adverse health effects and is not inherently unstable from a caustic, explosive or radioactive standpoint. The label in the upper left of the sheet, the one with no material on it, is a summation of all the most hazardous aspects of the others—it has the blue 3 from the KOH, the red 3 from the methanol and the yellow 2 from the KOH. This sign goes on your exterior shop door to inform anyone entering your workspace of the cumulative nature of the hazards within.

Use the links to the PDFs below—one for KOH, and the other for NaOH—and print out the signs in color and label the materials in your own shop. I use 8.5-by-11-inch white plastic label stock in my home color inkjet printer and then cut them out. Papilio white waterproof vinyl, 8.5-by-11-inch sheets are 10 for $12, available online. I cover them with clear packing tape so they last longer.

- troutsfarm.com/Biodiesel/4up_NFPA_KOH.pdf
- troutsfarm.com/Biodiesel/4up_NFPA_NaOH.pdf

Waste Glycerin

Waste glycerin from your reaction will contain most of the polar materials in your mix—methanol, glycerin, water, soap, residual caustic and some methyl esters. It is, therefore, somewhat flammable, toxic, caustic and not to be trifled with. It can be stored in HDPE plastic or steel containers, but should not be stored for long. It is also heavy—glycerin is denser than water—so containers over 5 gallons are hard to lift and carry without assistance. A 5-gallon container of glycerin bottoms will weigh over 40 pounds, and a 55-gallon drum will weigh over 500 pounds! I suggest determining the use or disposal option for your glycerin, and moving it out of your shop as often as you can, based on that. Don't stockpile glycerin—it is dangerous, messy and will alienate your friends, family and neighbors. This point underscores the need to include methanol recovery in your fuel-making operation. The only possible uses (known to me) for glycerol containing methanol are to feed anaerobic digesters and for combustion as an industrial fuel, both of which require a legally permitted facility to receive them. When you remove the methanol from your bottoms, other options are open to you—composting, dust suppression, adding to animal feed and making soap.

Used Wash Water

Used wash water will contain methanol and soaps—meaning it's high in biological oxygen demand (BOD) content, a measure of how polluted water is with material that will feed bacteria, algae and other critters that will pull out dissolved oxygen as they metabolize the nutrients. Used wash water should be left to settle to allow the biodiesel that it will invariably contain to rise to the top for collection and reuse. The water can be decanted off the bottom of your tank into another tank for pH balancing prior to disposal. Water pH should be neutralized with a weak acid—vinegar or diluted muriatic work well—to a pH of 7.0. If you perform methanol recovery on your methyl esters, settle and decant your wash water to remove any residual methyl esters and oil, then pH balance it. The used KOH in your treated wash water can be land applied and will act as a potassium fertilizer. Try this on non-critical plant life first!

Finished Biodiesel

Finished fuel can be stored in HDPE plastic containers, from the yellow plastic fuel jugs designed for petroleum diesel to 55-gallon poly drums to 270-gallon IBCs, where it is out of the direct sun and experiences very little daily temperature variation. One good rule of thumb is to not make more fuel than you will use in the next 30 days. Biodiesel should not be stored in steel 55-gallon drums as they are prone to rust, which clogs fuel filters and contributes to the early oxidation of your fuel.

If you follow these guidelines and are careful not to introduce particulates into your tank, you do not need a filter on your storage tank. In areas with high humidity, make sure the finished fuel container stays tightly closed to keep it dry. Every time you get to the bottom of your storage container, see if there's any sediment buildup—there shouldn't be with well-reacted, washed and dried fuel stored for 60 days or less.

Fire Prevention and Control

Fires caused by biodiesel production usually occur due to methanol or spontaneous combustion. Understanding how these fires occur can help you prevent them in your shop.

Methanol and its vapors are extremely flammable. Minimize the risk of fire by eliminating all sources of flames or sparks when methanol or its vapors are present. Eliminating release of methanol vapors throughout your process is a good place to start. Remember that open flames, light switches, thermostats and non-explosion-proof electric pumps are common sources of ignition in your shop.

In case of a methanol spill, remember that the fumes are toxic and their exposure is cumulative, so a number of small exposures build up in your system over time. If it's a large spill, you may be better off leaving the shop and getting a hose to water down the spill or even calling 911 rather than risk a significant exposure or possible explosion. For smaller spills, mixing four times or more water into the spill will greatly remove the fire risk, and then you can absorb small spills with Oil-Dri.

Spontaneous Combustion

Spontaneous combustion is the cause of many biodiesel-related fires. It occurs when materials like oily rags are left in a pile. Oxidation of the oils causes their temperature to rise, and due to their insulative nature, the heat builds up until the auto-ignition temperature is reached and a fire is the result. In addition to oily rags, spent filter material (like wood chips or Magnesol) is also prone to spontaneous combustion. I am personally aware of at least six of these kinds of fires in biodiesel plants in the US, told to me by the people involved. So it really happens and is not at all uncommon.

The way to minimize the risk of spontaneous combustion is to understand and never create the conditions under which it occurs. I don't use cotton rags for this reason. I use Oil-Dri on floor spills and wash tools and other equipment down with Simple Green and water. I do keep a roll of heavy-duty paper towels around as wipes but use them as a last resort and lay them flat between uses, and then throw them away when no longer useable.

If using shop rags, keep them in an airtight metal container to reduce the available oxygen and contain any fire that starts.

Fire Extinguishers

You should have at least one fire extinguisher, near the exit door in your shop, capable of extinguishing flammable solids, liquids and electrical fires. A 2A-10 ABC is a good choice for this purpose, available online for about $40. Additionally, you can install a heat-activated automatic fire extinguisher over your processor to activate in case a fire occurs when you're not in your shop. They can also be bought online for about $60 (search for Flame Defender 12 kg)—cheap insurance.

When to Call 911

It's important to know when to call 911 for help. My advice is if you think it might be a good idea, do it! Don't hesitate if a fire that looks beyond your control has started. If you are not in a position to call, yell for help and get that person to make the call. There is no need to be a hero, and thereby become a casualty. If you encounter a fire that appears beyond your control, move away to a safe distance and call 911. In the meantime, make sure no one else comes on the scene and assumes you're inside. Guard the site from a safe distance until help arrives.

A Valuable Resource

To close out this chapter on safety, I recommend an excellent resource created by my friend Matt Steiman, in conjunction with Pennsylvania State University, the "Biodiesel Safety and Best Management Practices for Small-Scale Noncommercial

Use and Production," published as a public service in 2007. I strongly suggest you download a copy and read it a few times as part of your ascent up the biodiesel learning curve. The URL for this PDF is in the Literature Review.

The Great Wall of Glycerin: Side Streams

The iconic image of backyard biodiesel is that of a Mason jar, usually backlit by the sun, in which two distinct phases are evident. The top phase is the beautiful amber liquid. That's biodiesel. That's what we came for. The bottom phase is typically a dark brown layer. If you have the recipe right, there is a clear line of separation between the two, and the two phases stay separated as you gently turn the jar one way or the other.

Biodiesel Cocktail

The bottom layer is a cocktail of products that drops out of a successful biodiesel reaction. As is, the cocktail is extremely difficult to get rid of. It's pretty much pure carbon, which means it has a value, but extracting that value is fiendishly complex.

The cocktail consists primarily of three parts: methanol, free fatty acids and glycerin. In it you will also find soaps and particulate. Because the cocktail is laced with methanol, it is a flammable material. That means that if you want to ship it over the road, it needs to be placarded and may need proper licensure. In North Carolina, you can legally carry 100 gallons of it on a normal driver's license.

Animals will eat free fatty acids, and glycerin, so an obvious target for the cocktail is animal feed. Because the presence of methanol diminishes its value as an animal feed, the cocktail needs to have what the USDA called GRAS Status, meaning "Generally Regarded as Safe." To achieve this status, it needs to have less than 140 parts per million of methanol. In other words, the methanol has to be removed from the cocktail before it will pass muster as animal feed.

You can compost the cocktail, but just as with the schmutz in the Grease Wars chapter, you need to be careful not to overwhelm your compost. When added to compost, the cocktail generates a lot of heat fast, so much heat that it can cause your pile to ignite. If you are going to try your hand at composting your cocktail, it is best to give it edges. That is, soak shredded paper, or wood chips or some other solid substrate so that air can get to liquid parts.

Anaerobic digestion is a perfect application for the biodiesel cocktail, but most backyard biodiesel operations don't have a digester handy.

At Piedmont Biofuels, we built a "bio-refinery" in order to sort the cocktail out. We built a flash evaporator, which we put under vacuum and heat. The methanol turns into a vapor at about 71.1°C (160°F), the vapor is sent to a condenser that is cooled by a chiller, and drip-by-drip, we can get our methanol back. That, of course, is an over-simplification. This recovery works on a sliding scale that is based on the percentage of methanol present in either the biodiesel phase or the glycerin phase, depending on which we are recovering methanol from. We do our methanol recovery under vacuum. At 26 inches of mercury with 5% methanol in the biodiesel, vaporization occurs at around 46°C (115°F). As the amount of methanol decreases, the boiling point rises. At 3%, the boiling point is around 54.4°C (130°F); at 1%, it is around 79.4°C (175°F).

Methanol

It is possible to do methanol recovery in the backyard, but it is seldom practiced. If you can get the methanol out of the cocktail, you will have a safer and much easier product to get rid of.

It should be noted that the methanol recovered from the glycerin cocktail tends to come back "wet." Moisture in feedstock, chemical water created in the reaction, and moisture taken from the air in the form of tank condensation ends up bound to the methanol, which is hydrophilic in nature. Wet methanol can be used when making biodiesel, but remember, "Water is not your friend," and it will diminish the yields that come from your process. Drying used methanol requires serious distillation capability. All of Piedmont's attempts to do this have failed—either from a lack of heat or a lack of capital to do it right.

Anyone capable of distilling water out of methanol in the backyard should probably bypass commodity products like fuel and turn their energy into those alcohols that humans like to drink.

Some municipal wastewater treatment systems will accept the cocktail as is. Methanol is a microbial starter—that is, communities of "bugs" explode in its presence—which means if you are trying to increase your population, it can be a welcome addition to the mix.

Municipal wastewater systems with digesters will often buy virgin methanol as feed for bugs. This can be true in college towns with vacillating populations. If your system is dependent on digestion, you need to keep microbial activity at a constant level. If all the students leave town, the volume of influent drops and there is a need to add supplemental feed.

The cocktail is sometimes used in spill remediation. Over the years, Piedmont has shipped hundreds of gallons into this specialty application. These clients tend to be environmental engineering firms who are hired to solve a brownfield problem. We once sold some glycerin cocktail to a firm that was remediating target ranges at Fort Bragg. Apparently bombs leave some sort of toxins in the soil, and somehow they figured out how to go after those toxins by land applying our glycerin cocktail.

Once the methanol has been removed, the glycerin can be used as a surfactant, that is, it can stretch water. Pure forms of glycerin are found in the soap that kids use to blow bubbles. Its surfactant qualities are what cause bubbles to hold together longer.

In periods of drought, people in the dust suppression business will use crude biodiesel glycerin in order to stretch their water supplies. Mines and other regulated entities that are required to keep particulate from their operations out of the air will buy crude biodiesel glycerin.

Another application for the biodiesel cocktail is as a "release agent." It's slippery, greasy and works great for getting things out of molds. Dump truck operators who are delivering asphalt to highway construction projects will sometimes spray it in their beds so that their entire product is released in a uniform fashion when they arrive at their target. That beats climbing into the bed with a shovel to get the errant chunks of asphalt that are left behind.

The coal industry is a big buyer of crude biodiesel glycerin. It helps them with their materials handling because glycerin, an alcohol, lowers the freezing point of water. Imagine giant piles of coal sitting in the rain. When the water freezes, the coal clumps together, making it harder to load onto rail cars and push into bins. So they spray the piles with the glycerin cocktail to reduce freezing and to facilitate handling. As a carbon product, the residue is seen as harmless when the coal is fired to make electricity.

The cocktail can be incinerated. Glycerin, methanol and free fatty acids are all legitimate sources of energy, but combustion needs to occur at extremely high temperatures in order to be complete. If you throw a glycerin ensconced log into your burn barrel, it will eventually ignite, and when it does, it will give off a deep black smoke filled with toxins and carcinogens of all sorts. Which means the biodiesel cocktail is not a good idea for fireplace logs.

Another problem with incineration is water. A great deal of the energy of the cocktail is consumed by driving off the water, leaving a poor return for heat. Piedmont once looked at burning the cocktail in its boilers and determined that the energy balance would be so poor it was not worth it.

Just like it is a good idea to develop a strategy for schmutz before you begin collecting used cooking oil, it is imperative that you develop a strategy for your cocktail before you go into production. And like schmutz, it will represent about 20 percent of your fuel production. If you make 1,000 gallons of fuel in a year, you are going to have 200 gallons of this pesky cocktail that will need a home.

I've seen homebrewers ignore this advice. They will fill up a 275-gallon tote with cocktail and put it behind the barn. As the years wear on, they end up with a great wall of the stuff. It's flammable. Stockpiling it is a good way to get closed down by the fire marshal.

Tote valves leak. Building the great wall of glycerin over time may also cause a spill. Cleaning up spilled biodiesel cocktail involves finding a new home for the dirt. It's not fun.

As always, the problem that plagues the backyard biodiesel enthusiast is scale. About everything that is traded is done so in full loads. Tanker loads of liquids are determined by weight. A tractor-trailer can haul 50,000 pounds of product, or about 7,500 gallons of biodiesel. Or 5,000 gallons of cocktail. To fill a tanker takes about 19 totes worth of product. That's a great wall of product.

Rather than trying to trade in the cocktail, find an outlet that can receive it in backyard quantities. If you have a community-scale biodiesel producer nearby, they will sometimes take it. Since Piedmont Biofuels is shot through with backyard sympathies, we receive it in 55-gallon drums and totes from smaller-scale producers all the time. But we don't pay for it. Our love of backyard production runs so deep, we plumbed our 10,000-gallon methanol tank in a way that we could easily fill 55-gallon drums for backyarders.

Wash Water

After you have found a home for the cocktail you will produce—either at your community-scale commercial fuel producer, or at your wastewater plant, or with a friendly engineering firm, or perhaps with a local asphalt hauler or perhaps with someone who needs to keep the dust down—the only other co-product of backyard biodiesel production is wash water. If you use water to wash your fuel, you will need a place to put it. The amount of water required to wash biodiesel varies widely with technology. The worst-case scenario is one gallon of water per gallon of fuel. Back to the 1,000-gallons-per-year model, this leaves 1,000 gallons of water to dispose of. As always, possibly too much for backyard composting.

Biodiesel wash water is a little bit like Jergens lotion: it tends to be white, thick and filled with soaps, trace amounts of methanol (not flammable) and glycerin. If you dump it on the lawn, it will kill the grass. If you dump it in the creek, it will kill all the living things in it.

Commercial composters tend to welcome biodiesel wash water. Some municipal wastewater treatment plants will accept it—it is high in biological oxygen demand (BOD)—and most facilities will charge to take in BOD by the pound. Substances that are high in BOD can pollute waterways. Think of BOD as food for algae. Dump your wash water in the creek and watch the algae bloom. When all the oxygen is consumed by the algae, everything else that makes its home there dies.

There are strategies for reducing the amount of water required for your backyard operation. They include water recycling—wash water clarifies with each successive washing, so the water used for the last wash can often be reused for the first wash. You can also use ion exchange resins. They will "polish" your fuel by extracting soaps and exchanging them for a free fatty acid.

Biodiesel production creates side streams. First and foremost is the glycerin cocktail, then wash water. There are also greasy plastic carboys, grease-saturated cardboard, endless biodiesel- and grease-ensconced rags, often pallets and plastic and metal drums.

The important thing to remember about side streams is not that they exist, but rather that you have a plan for proper disposal or reuse for each. Deal with them upfront by incorporating their existence into your plan, and your life as a homebrewer will be much more pleasant.

Regulations

For those who like to receive their good news first, it is perfectly legal to make your own fuel, put it in your own car and drive down the road. For those who feel no good work goes unpunished, the bad news is that biodiesel is a heavily regulated space.

People sometimes think that regulations surrounding food processing are arduous and expensive. Yet those are a walk in the park compared to fuel production rules. The government seems to care more about what you put in your tank than what you put in your mouth.

All regulations are mediated by scale, which means that backyard biodiesel production is much easier from a regulatory standpoint than is commercial production, but there are regulatory hurdles to address.

Grease Collection

Let's start with grease collection. Since all biodiesel production begins with feedstocks, this is the logical start point. Grease collection and handling has a patchwork quilt of regulations. Different states—even different municipalities—have diverse requirements for collecting and hauling used cooking oil.

In North Carolina, you need to be able to prove chain of custody for the grease in your possession, and you need to have a one-million-dollar liability policy in place. Co-op members, backyard brewers, small producers and biodiesel enthusiasts fought against what became known as the "Grease Police" bill for years. Here's a blog entry that I wrote when it appeared we were going to lose the fight.

Grease Police

I was bludgeoned at the North Carolina legislature today. It looks like "Big Fat" is going to win this state, just like they did in Virginia.

For the past couple of years we have been in a pitched battle against what has become known as the "Grease Police" bill (HB512).

I have to admit I am stunned that its Republican sponsor John Torbett from Gaston, and Jimmy Dixon from Duplin/Onslow are introducing legislation that increases the regulatory burden on small businesses. Wow.

Let's do this. Let's run for office on the small-government/anti-regulation platform. Let's win both the NC House and the NC Senate for the first time since the Civil War. And once we are in power?

That's simple. Let's regulate dozens of small grease collecting businesses out of existence with new licensure requirements, new reporting requirements, new inspection regimens and new labeling rules. Got it.

What astonishes me the most is that none of them seem to see the hypocrisy in what they are doing.

A Raleigh law firm on behalf of Valley Proteins wrote HB512. They are the giant Virginia-based rendering company that operates in North Carolina as Carolina By-Products. With 1,400 employees, they have some reach in our neck of the woods. Allegedly the "North Carolina Renderers Association" has asked for increased regulation of itself. I'm not sure who they are—other than a P.O. Box in Virginia. I think it would make it simpler if we just referred to them as "Big Fat."

What I know about passing legislation you could put in a thimble. Today was like walking to a schoolyard fight without anyone at my side. I showed up to a room full of lawyers and lobbyists who made it look like the passing of this bill was a forgone conclusion.

The chairman of the subcommittee was Mike Stone from Harnett/Lee. He seemed to get it. Among other things he seemed to understand the restaurant business. He's a business guy from Sanford (a town where we collect used cooking oil), and he seemed to understand that the restaurant operator needs a service provider that can respond to their needs.

He ran the meeting like a Swiss train conductor, giving all sides a chance to talk, and staying on time. You would think that passing a bunch of new regulations on small business would leave a bad taste in his mouth, but I guess it is hard to stare down a bunch of highly paid lawyers who are retained by Big Fat.

Also on the subcommittee was Darrel McCormick from Iredell/Surry/Yadkin. He seemed more informed and more relaxed than the others. He's also a business guy, and he did have a moment of clarity in which he questioned why they were even wasting their time on this bill. I don't think his nature is to increase regulation on small business, but then again, he seemed to think it was inevitable.

Joe Hackney from Chatham/Orange was also there. He's the ex-Speaker of the House who completely gets it. He suggested the bill be thrown out, and I will say that both McCormick and Stone seemed to treat him with deference and respect.

Torbett had nothing of value to add to the conversation, except that used cooking oil had become a commodity.

Duh. That being the case, let's pass a law that some lawyer has written for us that will put a bunch of small-scale grease collectors out of business. In his own county they used to collect used cooking oil, spin it into biodiesel and use it to power their school buses.

I can see shutting that down.

Best to import more fuel from the Middle East.

By the way, I should note that if HB512 becomes law, it will not affect Piedmont. We will buy our license, or licenses, as needed, and we will conform to the reporting requirements, and we will comply with the new labeling, and we can afford the insurance they demand. It won't put us under. It will simply drive up our operating costs, which we can happily pass along to the folks who use our fuel.

It's not a big deal really. It's just more government. And the increases are only pennies per gallon, really.

Apparently the North Carolina Department of Agriculture will be the agency tasked with doing the inspections. They did voice concern that the licensing fees might not cover the manpower required to comply with this bill.

But that's OK. Once it is passed into law, you can always just jack the fees to pay for the bureaucrats to administer the new regulations.

It's unbelievable. Today I had a front row seat in how big government is created.

I should say that Big Fat has couched this whole debate in the context of "Grease theft." They estimate that they have lost millions of dollars over the years since biodiesel came along.

What they fail to understand is that we are not stealing their grease. We are simply stealing their accounts. Restaurants sign up with grease collectors based on the best service, and the best value and the best deal for them. Today Big Fat does not offer the best deal. Which means they have been losing market share.

Valley Proteins took one of Piedmont's grease collectors to court. She was charged with stealing grease. When she brought the restaurateur into the courtroom, along with the contract they had signed, the judge threw the case out.

Let's see. Big Fat can't seem to win in the marketplace. And they can't win in court.

So they have decided to go win down at the State House. All they need to do is pass a law that increases the regulatory burden on small-scale grease collectors to the point of extinction, and bingo, they can regain their market share.

And what is astonishing is that the Republicans are letting it happen.

North Carolina's economy will be diminished. Virginia's balance sheet will be enhanced. And no one seems to have any trouble with this picture.

Oh well.

We tried. Since we can't afford a lawyer to sit outside of a representative's office all day, like Big Fat can, and since we are simply a small business trying to find our way in the world, I guess we will just comply.

Let me find the stencils. I'll start re-labeling the trucks myself...

The bill was the brainchild of Valley Proteins and the North Carolina Renderers Association, and what started out as a significant overreach—demanding licensure, truck labeling and annual inspections in a misguided attempt to make biodiesel companies "renderers"—ended up severely diluted. It did make the theft of used cooking oil a felony, however.

If feedstock is the place we begin, don't steal it. Come up with a simple contract, and get the establishments supplying the oil to sign it. Check with your town and your state to see what other grease collection regulations might be brought to bear. On the following page is an example of the Oil Removal Contract that Piedmont uses.

Methanol

The next thing you'll need is methanol. In North Carolina, it is legal to carry 100 gallons of methanol on your truck without special licensure, or placarding requirements. Using the 20 percent rule of thumb, each methanol run can provide enough reactant to knock out 500 gallons of biodiesel.

It's relatively easy to be legal and compliant when collecting your oil, your catalyst and your methanol. The next trick is to determine where you are going to make fuel. Local zoning can come into play here. Most residential developments will frown on backyard biodiesel operations—but many will have no rules or guidelines against it. Farms are good. Any time you can locate your operation on a farm, you will enjoy huge regulatory benefits.

While the planning department of your locality will determine the zoning of your operation, the most important actor is your local fire marshal. Fire marshals have a lot of discretion in interpreting universal fire codes. A good relationship with your fire marshal is your most important regulatory start point.

Fryer Oil Removal Contract

Date: 10/17/14   **Location:** Local Restaurant with address
Primary Contact: Name, cell phone, land line, email address
Container Type: 55-Gallon Drum
Frequency of Pickup: As needed
Number of Fryers:

This agreement between Piedmont Biofuels and _____ gives Piedmont permission to locate a used fryer oil collection vessel at the above service location with the exclusive right to pick up oil from said vessel. By signing this agreement, the customer and Piedmont agree to the following terms and conditions:

The customer will provide a suitable location for a used cooking oil collection vessel and will deposit all of its used cooking oil in that vessel. The customer will not use any other company for waste oil collections or have any other used cooking oil vessels present at the location.

Piedmont Biofuels has permission to come on the property as often as is necessary to service the vessel.

Piedmont will be mindful to keep the area neat and clean, and to maintain and empty the vessel(s) in a professional and work-manlike manner. The customer agrees to be mindful of its used cooking oil disposal, to keep lids in closed positions and to keep the vessel area clean and neat.

Piedmont has the exclusive right to collect all of the used fryer oil from this location.

This contract is valid for 3 years, and will continue on a month-by-month basis at the end of that time.

Piedmont agrees to pay _____ for the duration of the contract, increasing or decreasing the rate periodically based on the market rate for yellow grease.

By signing this agreement, the customer agrees to terminate all other oil collection agreements at this location and agrees to notify other oil collection companies engaged at this location to remove their property within 14 days of the contract date above.

Signed,

_____
Authorized Customer Piedmont Biofuels

Before you run in fear from your fire marshal, check him or her out. There's a good chance they will be as interested in biodiesel as you are. If you are clear about your scale and invite them to visit your facility—so they know what is going on—they can be extremely easy to work with.

Fire marshals are also able to dictate spill containment requirements, and building spill containment into your facility is an excellent idea. The general rule is that a spill containment area should be able to hold 110% of the largest vessel in the containment.

One caution is that you will need to have your methanol storage, your methoxide mixing and your biodiesel reactor set up outdoors—away from nearby buildings. Indoor methanol (which would include your glycerin cocktail) requires a sprinkler system. Washing, drying and storing biodiesel does not require a sprinkled facility and can be done indoors.

Wash Water

Wash water is another area of your operation that may or may not be dependent on regulations. North Carolina has a primitive wastewater regulatory regime such that wash water cannot be land applied. You would probably not want it in a septic system, but depending on volume, it may be something that is readily accepted by your town's sewer authority.

Glycerin

The last piece of the equation is the glycerin that is created from your biodiesel production. It needs to be treated as a hazardous material as long as it contains methanol. If you have built methanol recovery into your plans, you may be able to produce a glycerin that is not hazardous for transport. Recall that 100 gallons is the magic number in North Carolina. Check with your Department of Motor Vehicles to see about your local regulations.

Taxes

If you are producing fuel only for your own consumption, you are ready to fill up and drive. Now you need to consider taxes. Every gallon of biodiesel in America is subject to state and federal road and excise taxes. Recall that in North Carolina the first 2,000 gallons of output is free of North Carolina road tax. Federal remittance is still required. Your Department of Revenue and the IRS will be delighted to help you figure out how to remit your taxes.

I think it is worth mentioning that many folks are drawn to backyard biodiesel in an effort to evade taxes. In North Carolina, there is $0.62 of tax in every gallon—which is a pretty good size chunk of a $4 fuel.

Not paying taxes is a moral decision. Those who wish to withhold the payment of taxes on moral grounds—pacifists, for instance, who do not want to see their taxes spent on military endeavors—are free to do so and take their chances. But road taxes are different. If you don't want to pay them, simple— Don't use the road.

Some of our regulations are designed to keep our property and our bodies safe from harm. Some are designed to keep pollution out of the environment. Pollution is a function of scale. If you spill some grease on the grass, it will turn black and die. Over time, the soil will metabolize that grease, and new grass will grow again. By the same token, if you spill a truckload of mother's milk in the creek, you can kill everything in the water. That means that being mindful of your scale is important. It's one thing to accidentally knock over a bucket; it's another matter to lose an entire 275-gallon tote of liquid due to a leaky valve.

Myths and Realities

I started running on B100 biodiesel in 2002, and over the years, I have seen a number of issues arise—some real, and some imaginary.

Biodiesel as a Solvent

At 100% strength, biodiesel is a powerful solvent. That's not a myth. It will devour rubber components of fuel systems, it will strip the paint off of your car, it will eat concrete, and it will dissolve pavement. Vehicles manufactured before 1994 can have rubber components in their fuel system, and users who switch to B100 will find those components destroyed. It's not a question of "if"—it's a question of "when." Rubber components can be as simple as a return line—which is easy to replace with either Viton or a less expensive urethane line. Seals in injector pumps are more complicated. An easy way to avoid problems with rubber is to procure a post-1994 vehicle.

Also, biodiesel does not get along with copper. The two react. The unfortunate thing about that is that many fuel gauges are dependent on a submerged "sending unit" that has copper components. That means B100 is a great way to wreck your fuel gauge—but that's OK, if you are making your own fuel, you will have an excellent idea of how much is in your tank at all times.

Engines that have a lot of miles driven with petroleum diesel can accumulate carbon deposits, and the solvency aspect of B100 tends to have a "cleansing effect." Switching directly to B100 can clean out residual carbon, and that can occasionally clog fuel filters. This filter-clogging effect is temporary, and once the fuel system is cleansed, the filter plugging stops. I should note that this is not guaranteed to happen. In our Piedmont Biofuels Membership Agreement, we suggest that new members throw a fuel filter in their glove box—and know how to change it during their transition to B100. I have bought several high-mileage diesel vehicles over the past decade and not experienced any filter-plugging issues—and I have seen issues arise with members after a single fill.

Initial filter plugging is real, albeit temporary, and not guaranteed. Some users report their engines run a little quieter on B100, after the carbon deposits have been cleared. Others report higher miles per gallon on biodiesel, a claim I believe to be mythical.

B100 and Pumpe Düse

In 2006, Volkswagen America switched its Turbo Direct Injection (TDI) models from common rail to pumpe düse. That was the first time that it was supposed to be the end of the B100 driving community. That engine technology change got a lot of buzz in the B100 world, but the idea that B100 would not work with pumpe düse turned out to be a myth.

Biodiesel and Cold Weather

Cold weather can certainly be a challenge for the B100 enthusiast. In cold temperatures, petroleum diesel has a tendency to gel. That's why you will often find block heaters on diesel motors. Gelling is a much greater problem for biodiesel than it is for petroleum diesel. When biodiesel gels, it cannot be pumped, and it cannot pass through fuel filters, so the diesel engine will sputter and die on gelled fuel. More likely, the engine will refuse to start in the first place.

All biodiesel has a gel point—that temperature in which it changes from a liquid to a solid, and once a solid, all biodiesel will return to a liquid state when it is above its gel point. When I have gelled fuel, I typically push my car into the sun, call the Plant and tell them I will be in late. Cold Carolina mornings tend to become warm Carolina afternoons, and gelling is not nearly the problem for us as it might be in colder climes.

The temperature at which biodiesel gels is directly connected to what feedstock was used. Fuel made from animal fats will gel at a higher temperature than fuel made from vegetable sources. A handy way to think of this is to fry up some bacon and fry up some potatoes in vegetable oil. Have a great meal, collect the drippings from each and put them on the windowsill for the night. In the morning, the bacon fat will be a solid, and the vegetable oil will still be flowing nicely. Those same characteristics pass through into fuel made from each.

This is also true of varying plant feedstocks. Fuel made from soybean oil (a warm-weather crop from the Midwest) gels at a higher temperature than fuel made from canola (a cold-weather crop from the north).

Over the years, I have seen a variety of strategies to fend off cold-weather gelling problems—straw bale passive solar garages, inline heaters, electric blankets, block heaters, the works—and the best remedy is good old-fashioned petroleum diesel. Cutting B100 with petroleum diesel lowers the gel point dramatically. I used to brag that we could get through any North Carolina winter on a B80 blend—by cutting our fuel with 20% petroleum—but now I am not so sure. In the winter of 2014, we had polar vortex after polar vortex that sent me to the gas station to increase my petroleum rations.

At Piedmont Biofuels, we buy a load of bulk petroleum around first frost, we tell our members that the blending time of year is upon us, and the B100 Community Trail that we operate becomes a misnomer. We pump B80 until the weather warms up.

Of the seven stations on the B100 Community Trail, we leave one as B100 year-round—to placate those members that want to do their own blending, or that have their own cold-weather strategies. These folks tend to be the ones you will see on frosty mornings dumping tea kettles of boiling water onto their fuel lines, or crawling beneath their vehicles to apply hair dryers to their fuel tanks. A word of caution: fuel tanks on Volkswagen TDIs are plastic and can be melted clear through with a hair dryer.

The best strategy? A few months of petroleum.

Biodiesel and Warranties

Vehicle warranties are another issue that has long plagued the B100 community. In the early days, we used to wave the *Magnuson-Moss Warranty Act* around, a law passed in 1975 that stipulated manufacturers could only warrant parts and labor. Our logic was that if you can only warrant parts and labor, you can't void an engine warranty by what you put in your fuel tank.

And this logic does hold true. If you lend your brand-new diesel Mercedes to your newly licensed teenager for prom night, and they accidentally fill it up with gasoline instead of petroleum diesel, you can tow it to the dealership and find that you are not covered under warranty. It's also true of "bad gas." When you are low on gasoline and use some two-stroke oil and gasoline mix from the tool shed to get to town for a fill, you will find you don't make it to town, and you are not covered under warranty.

The "warranty" issue has moved around a lot. Volkswagen America is notoriously bad about this. When my wife, Tami, bought a new Jetta Wagon in 2003, she drove it out of the dealership from beneath a banner that read, "Biodiesel will void your warranty." She did need to get the radio replaced during her warranty period, which I believe the dealer blamed on biodiesel. After three rear-end collisions, and 283,000 miles on B100, Tami traded her Jetta in on a 2012 Beetle.

It's interesting to note that Volkswagen Europe warrants high-percentage blends of biodiesel and Volkswagen America does not. When you ask the folks at Volkswagen about this, they tell you it's because of different iodine values in each continent's respective fuel. In Europe, a lot of biodiesel is made from rapeseed (which we call canola). By all accounts, the notion of iodine value having a relationship to engine warranty is utterly silly, and Volkswagen's hypocrisy on the warranty issue is manifest in North America where cars on the eastern seaboard are only "warranted" to run B5, yet the same makes and models in Iowa (where biodiesel is prevalent) are warranted to run B20.

My favorite warranty story comes from the Cincinnati, home to Peter Cremer's biodiesel plant. He is legendary for having produced the first biodiesel in America—throwing reactants in a rail car and shipping it to Colorado, and telling their customer to ship the glycerin back.

The City of Cincinnati had a positive experience running on B20 and decided to move its bus fleet to a B50 blend. About the same time, it put out a contract for additional buses. Detroit Alison won the bid, shipped the busses to Cincinnati, and when they learned the buses were to be run on B50, they claimed that was impossible because it would void the warranty. When the City of Cincinnati replied that, if that was the case, they could come and pick their buses up. At which point, it was OK to run B50 in Detroit Alison engines.

While warranty issues have been a Wild West for B100 users, new vehicles are coming out with warranties that expressly forbid high-percentage blends of biodiesel. Piedmont Biofuels has one such model—a 2014 Chevy Cruze. It's been on B100 all its life, and as such, its warranty is void. The vehicle runs fine, although the check engine light comes on. When we take it in for diagnostics, the dealer is unable to identify the problem—although surely it is B100 that is causing the problem. In the case of the Chevy Cruze, Bose makes its electronic brain, and a German firm reprograms the Bose brain such that the check engine light will not come on when running B100. Again, Europe is ahead of North America in its adoption of B100.

If warranty is a stumbling block, that's simple: buy a car without one, and it will run fine.

Biodiesel and Diesel Particulate Filters

It's impossible to tackle biodiesel myths and realities without delving into diesel particulate filters (DPFs). This relatively recent technology change has led to what are commonly referred to as "clean diesels."

No one can argue that cleaning up diesels is a good thing. Whenever you are behind an old poorly maintained dump truck at a stop light, when its acceleration fills the street corner with thick black smoke, diesels get a bad name. Consumer choice for diesel vehicles in America is a fraction of what it is in Europe, and some say this can be tracked back to an anti-diesel bias in the US Environmental Protection Agency.

Clean diesels, thanks to DPFs, are changing this. When once the biodiesel enthusiast had predominately only Mercedes or Volkswagen passenger vehicles to choose from, there are now clean diesels from General Motors and Audi entering the American market.

DPF technology varies widely in its deployment, and it has rapidly become the bane of diesel mechanics everywhere. Some have rechargeable urea-based systems. Others backwash filters with fuel and deposit that spent particulate-laden fuel into the crankcase oil system.

The introduction of DPFs is another moment in B100 history where it is allegedly "over" for the use of the fuel. It turns out that too is a myth. One of the popular misconceptions surrounding DPFs is that using B100 will cause engine oil dilution, which will increase wear, which will wreck your engine.

At Piedmont Biofuels, we have many members running newer vehicles with DPFs on B100. We've gathered crankcase oil samples at scheduled oil change intervals, and we've searched for dangerous dilution rates, but we haven't been able to find them.

A great disservice to the B100 community was done by biodieselSMARTER, when it published a piece by Jason Burroughs on a case of "bio-indigestion." With no science or repeatability, sheer speculation linked poor engine performance to a DPF and B100 usage. As can be the case in the Internet age, the article went viral, was picked up by *Popular Mechanics*, among others, and laid an inaccurate cornerstone for the B100 DPF conversation.

The question of whether or not DPFs are a problem for the B100 community remains open. We have many members with clean diesels, we have DPFs in our own fleet running on B100, and we have many miles with no problems to report. That said, DPFs have imposed a burden on diesel operations in general, and B100 is not immune.

Biodiesel and Exhaust Smell

No discussion of biodiesel myths and realities would be complete without mentioning the smell of exhaust. Just as the cold-flow characteristics of feedstock pass through into fuel, so does aroma. If you have heard that biodiesel exhaust smells like French fries, that can be true. Piedmont Biofuels is a "multi-feedstock" producer, which means a variety of feed-stocks passes through out plant. Over the years, I have had exhaust that smells like French fries, and dog food, and I have had exhaust that smells like lemon-infused almond oil. Also mint, when we procure waste grease from makeup companies. The worst exhaust smell by far is the one derived from algae oil. That's when I drive down the road smelling like spoiled fish.

The Community

My favorite aspect of homemade biodiesel is the community of people it attracts. This seems to be a lesson that we get to learn over and over again. I reflect more broadly on this idea in *Industrial Evolution*, but here's a shorter version.

When you are making fuel, it is easy to get lost in the technical details. You think it is about pumps, and chemistry and energy balance, when in fact they all have very little to do with it. Making fuel is about the people who want to drive around on that fuel.

At Piedmont, we find this is true of everything. When you are creating a farm, you think it is about fences to ward off the deer pressure, or about bed structure, or about irrigation strategies, when in fact it is about the people who want to eat the food you are growing.

I think this is true of the solar energy community as well. At the plant, we have a 94 kWh solar rig in the yard. Because it is above the north field of Piedmont Biofarm, we refer to it as "solar double cropping." When creating a project like this, it is easy to get lost in the details of interconnection agreements

and soil compaction and how many electrons will be generated, but it is really about those people who value clean solar
electricity.

Cheap Fuel Folks

The backyard biodiesel community is characterized by distinct
groups of individuals—and I am going to stereotype here, because stereotyping is one of the things I do best. At its heart
is the "cheap fuel folks." These are the farmers, and loggers and
truck drivers who are desperately seeking relief from high fuel
costs, and they tend to burn a lot of fuel. When fuel costs have
a material impact on your monthly budget, it is easy to turn to
homebrewing fuel as an escape route.

The cheap fuel crowd tends to be rabid do-it-yourself enthusiasts. They can often do their own plumbing, welding and
rigging. And they are frequently shade tree mechanics. Often
they start out life as diesel enthusiasts, so the road to homemade biodiesel can be a short one.

Being inept at auto repairs and maintenance, I am not cut
from this cloth, although I have been known to rig up a few of
my own inventions from time to time.

Do-it-yourselfers delight in the inventions of others, and
like showing off their own work. At the same time, a lot of
accomplished homebrewers remain "underground." Those
who don't comply with zoning, fire or taxation regimens tend
to quietly make their own fuel without attracting attention to
their projects.

Very early in the formation of Piedmont Biofuels, we decided to stay as legal as possible so that we could tell the world
about what we were doing. We wrote blog entries and books
and posted to forums and spoke at conferences and civic organizations—shooting our mouths off wherever possible. In
some ways, I feel Piedmont cleared some space for the rest of

the community to inhabit. When we started homebrewing in 2002, for instance, the State of North Carolina required anyone making fuel to post a $2,000 bond with the Department of Revenue's Division of Motor Fuels in order to cover potential tax liability. We correctly explained that most homebrewers didn't even invest $2,000 in their entire setup and would be extremely unlikely to post such a bond.

We posted one for Piedmont. To keep ourselves legal. We then proceeded to remit fuel taxes for anyone in the state. All that was required was that they join our co-op (at a cost of $50 a year) and send their fuel tax money to us, so that we could remit on their behalf. It was as if their gallons were our gallons, and it kept everyone legal.

It also gave me a chance to publicly bash Norris Tolson, then Secretary of Revenue. Time and again, we explained that our antiquated taxation paradigm was forcing homebrewers to stay underground. Through our endeavors, they changed the law such that anyone making their own fuel was entitled to a fuel tax exemption on the first 2,000 gallons produced.

Norris and I went on to become friends, by the way, and together we were co-founders of the Biofuels Center of North Carolina.

Frequently you find the cheap fuel crowd overlaps with other energy cost-saving strategies—like pellet stoves or owner-built homes. In the grassroots biodiesel movement, these folks are occasionally referred to pejoratively as "tire burners," implying that they would heat their homes by burning tires if they could. These folks cannot comprehend why anyone would want to pay more for biodiesel than petroleum diesel, ever.

Clean Air Folks

Another distinct group in the biodiesel community are the "clean air folks." Many work in policy roles, but they drive around on biodiesel and promote biodiesel because of its superior emissions profile. The clean air cohort tend to be evangelists—staging clean driving technology conferences and often promoting electricity, natural gas and carpooling as alternative transportation. These are the bike riders, and the "smart" commuters in the lot.

These folks are less likely to make their own fuel, but readily pay a premium to fill up on B100 biodiesel. They have a clear understanding of how fossil fuels have externalized their true costs and are grateful to have biodiesel available in their community. In general terms, whenever air quality is in decline, biodiesel consumption is on the rise—air quality is the driving force behind much of biodiesel sales.

Other Biodiesel Community Folks

Highly visible portions of the biodiesel community are those who are drawn to the fuel to escape the geo-political aspects of the oil business. This is the "No War Required" crowd and tend to have the loudest bumper stickers.

By the time you throw in the tree huggers, those environmentalists who cannot abide dead pelicans on the Gulf Coast, a picture of the community is complete. I frequently tell my tour guests that some of Piedmont's members are far-right-winged survivalist nuts who want to pay for their fuel in constitution silver, and that they stand side by side with the hippie chicks who want to trade fuel for massages.

What's interesting about the community is that its diversity gives it strength—just as diversity in nature strengthens ecosystems. When you are trying to fend off over-reaching regulations on grease collections, ironically passed by North Carolina's Tea Party legislature, it's helpful to have Tea Party enthusiasts in the mix. And when you are trying to change tax

laws, in order to give biodiesel preferential status, it's handy to
have believers in government at your side.

Beyond North Carolina, there is a close-knit community
of homebrewers, small cooperative producers and community-
scale biodiesel plant operators. This community rises and falls,
like anything else, but it tends to be permeated by deep believers who preserve against great odds.

Biodiesel Online Community

Biodiesel also enjoys a remarkable online community. Its arrival
on the doorstep of America's fuel mix coincides nicely with the
dawning of the Internet. Since many backyard enthusiasts are
also digital natives, there has long been a robust dialogue of
various topics—ranging from pricing to technology to new analytical approaches (and everything else for that matter) in the
form of forums, blogs, list-servers and websites.

One of the busiest and longest-lived ongoing forums is
biodieseldiscussion.com that a started over 15 years ago. It is
a fantastic place to lurk and learn, or to dive in and share with
the community. It's also a place to forge professional as well as
personal connections with like-minded individuals. New fuel
makers should be aware of two in particular as developers of
resources for them.

RickDaTech (aka Rick Boggan), who has been part of
the biodiesel community for over 10 years, started a biodiesel
homebrew supply store, B100 Supply, in 2005 to meet the
needs of enthusiasts nationwide. Two of his accomplishments
merit mention.

Foremost, Rick created and maintains one of the best com-
pilations of biodiesel homebrewing information on the net.
He built this site after gathering information from the biodiesel discussion forum as well as soliciting input from the

community. He added his own experience, and the result is an accessible yet comprehensive library of topics on homebrew biodiesel. His other significant contribution is a great manual on building and using an Appleseed processor, which I highly recommend if you intend to make an Appleseed the basis for your fuel making. The book is for sale on Rick's website, make-biodiesel.org.

Graydon Blair, founder of Utah Biodiesel Supply, also deserves mention. In addition to continuing to sell all manner of biodiesel equipment, reagents, lab ware and the like, he has produced instructional videos to help biodiesel homebrewers visualize everything from titrations to measuring methanol purity. He also started his business in 2005, and along with Rick has been a regular contributor to discussions on the Biodiesel Discussion forum. His other major contribution as a constant cheerleader for biodiesel was manifested in his role behind the Collective Biodiesel Conference that started in Golden, Colorado, in 2006. He has helped make holding the event on an annual basis possible. It has become a get-together of biodiesel homebrewers, grease collectors, educators and small commercial producers that is always worth the price of admission.

Biodiesel Organizations and Conferences

Organizations in the community tend to be bootstrapped and remarkably resilient. One of them, the Sustainable Biodiesel Alliance (SBA), is anchored by the folks at Pacific Biodiesel in Hawaii and is frequently populated by movie stars and rock stars. It can largely be credited with getting biodiesel into the tour buses of folks like Jack Johnson, Willie Nelson and Dave Matthews. Small operators across the country (including Piedmont) have benefited mightily from playing an occasional role as "fuel attendants to the stars," having the opportunity to fill up tour buses and transport trucks of big traveling shows when they pass through our communities.

The Sustainable Biodiesel Summit (SBS) began by "shadowing" the National Biodiesel Board (NBB) annual conferences—in effect, holding their own self-contained events where small producers could have an effective voice and could learn from one another. It can be credited with the creation of the "Small Producer" category of the NBB. At first, the NBB imposed a volumetric dues collection scheme on all of its members, but thanks in part to pressure from the SBS, they created this category so that small producers could participate as members with a flat rate. The SBS has attracted remarkable talent to its conferences, and at its height, it rivaled the NBB for valuable content. Richard Heinberg, Jon Van Gerpen, Josh Tickell and indeed most of the authors cited in the Literature Review chapter have addressed meetings of the Sustainable Biodiesel Summit. For years, it was attended by the who's who in grassroots biodiesel production.

Tensions between the industry's trade association and Joe Jobe, the CEO of the National Biodiesel Board who frequently attended SBS to address small producer concerns, often ameliorated the SBS. The strength and presence of the SBS caused the National Biodiesel Board to dedicate some staff resources to the subject of sustainability. Under the leadership of Don Scott, this is now on the radar and a frequent topic at NBB conferences.

While the SBA was hobnobbing with celebrities, and the SBS was grinding out shadow conferences, a third grassroots biodiesel event emerged, the Collective Biodiesel Conference, (CBC) which was rooted in geek nature. Less interested in policy gains, and more interested in pumps, tanks and better backyard production methods, this conference that arose in Colorado has been staged across Canada and the United States.

Bob Armantrout made his big community debut at a CBC conference at the Colorado School of Mines in Golden. His Hawaiian shirt and mastery of gas chromatography (GC) left an indelible mark on conference participants. At a time when fuel quality was at the industry's fore, Bob demonstrated how the GC could play an important role, and how it could be accessible as a device to small-scale producers.

In 2011, the CBC took a fascinating turn. Starved for resources, and a labor of love, it chronically teetered on the brink of extinction. When it was picked up by the folks at the Cowichan Bio-Diesel Co-operative on Vancouver Island, British Columbia, it was given new life, and since then, it has become a conference that people have to more or less "bid" on to bring to their community.

A volunteer board of small producers now accepts applications from around the continent and awards the conference to the most competitive proposal. This appears to be a masterstroke in that the CBC has remained healthy ever since.

In 2014, the CBC was awarded to Piedmont Biofuels and brought to our southern town of Pittsboro, North Carolina. In collaboration with Central Carolina Community College, Abundance, North Carolina, and the Center for Democratic Communities, we staged a remarkable show.

Bob ran the conference like a Swiss train conductor—with each session beginning and ending on time. Great care was taken to ensure excellent local food was served at every meal, and a star-studded cast of speakers filled the schedule. Conference participants attended sessions at Central Carolina Community College by day and social events and tours at Piedmont Biofuels at night.

As someone who had a front-row seat in the planning and implementation of this conference, I was moved by the deep love and understanding that permeated this group of people. Policy makers, backyarders, small producers, grease collectors, vendors, technology providers, researchers and biodiesel consumers all attended this event. While many of the relationships were new, others went back many years, and the camaraderie and esprit du corps were impossible to deny.

Bob kicked off the 2014 Collective Biodiesel Conference with his talk "What a Long Strange Trip It Has Been," complete with psychedelic music and some face tugging in a simulated acid trip. By all accounts, this was one of the conference highlights—but I took umbrage with his "Six Steps of Biodiesel Addiction." Weeks later, I did a canoe trip in Quebec with my son, Arlo, and I attempted to superimpose my canoeing impressions over Bob's opening remarks. I published "Stages of Grief vs. Riding the Rapids" in Energy Blog.

Stages of Grief Vs. Riding the Rapids

Bob has Stage One as "Love at First Sight." And I agree. When people first encounter the notion of driving around on fuel made from waste vegetable oil, their imaginations ignite. It's kinda like "Day One" on a canoe trip. Your gear is dry, your body is strong, you are well rested and well fed, and the river is downright inviting.

Stage Two is the "Honeymoon," according to Bob. This is when you are evangelical. Bragging, even. This is when you think, "I'm in an open canoe, happily paddling down this river—look at me—life is good."

Stage Three is "Nest Building." That's when you get into a diesel vehicle, acquire equipment and start collecting grease. This is the stage when you get "free" of the petroleum grid. This is when you pitch your tent on a sandy beach and cook dinner on a campfire beneath the stars.

"Maturity" is the word Bob uses for Stage Four. This is when you learn to clean up spills, and deal with cold weather and inconvenience. This is when you learn that making your own fuel is challenging.

In "canoe" terms there are "swifts," and rapids on the river. The rapids are classed from one to five. Open canoes loaded with hundreds of pounds of gear can only handle Class Three rapids. Others need to be portaged around. Once you have shot rapids, swifts are just fun.

There are rapids in maturity. You hit rocks, the boat spins, you take on water, and your gear gets soaked. The river becomes a heartless mistress.

Bob has deemed Stage Five "Disillusionment." In biodiesel terms, this includes the fires, and grease wars and inability to make any money. In canoe trip terms, this is when your boat swamps or gets pinned to a rock. This is when you are bug-bitten, sunburnt, and your shoulders burn from portaging with a heavy canoe. After a few days of sleeping on a bag of wet clothes, on the hard ground, your body aches, and you don't know why you are on this forsaken river in the first place.

Stage 6 is "Divestment." For Bob, this is an advertisement on Craigslist. You shed your gear, and perhaps your biodiesel car, and you walk away from the whole ordeal.

Piedmont routinely benefits from this divestment stage. We are often called upon to pick up abandoned feedstocks, or great walls of glycerin co-products. We sometimes walk away with fittings, or static mixers or tanks that are shed during the divestment stage. If you can stay standing long enough, a blue bird occasionally comes your way.

This is where the canoe trip analogy departs from the Stages of Grief.

After days on the river, your skills have increased. You can handle larger rapids. You can steer around rocks. You can avoid standing waves that might fill your boat with water. And you can artfully dodge "holes" in the river that can overturn your canoe.

Arlo and I began our canoe trip as the inexperienced boat in the lot. By the end of our trip, we had developed excellent communication between the bow and the stern, and we had shot a Class Three rapid, fully loaded with wet gear. We were routinely leading our companions through unknown fast-flowing waters.

Instead of divesting, we were jazzed. Bruised, battered, soaked and exhausted, we were invigorated and ready to do it again.

2014 is not a very good year to be in the biodiesel business. Our production credit lapsed, and the petroleum interests are winning on Capitol Hill. The Renewable Fuel Standard which fuels the Renewable Identification Numbers (RINS) market is under assault, and it's just not a pretty time to be a commercial producer.

Piedmont has traveled through all of the stages Bob outlined, except divestment. We've had accidents and spills and fires, we've made money and lost money in this business, and instead of divesting, we are ready to take our skills and paddle out onto the next river.

Bring it....

Biodiesel makes for strange bedfellows. Oilmen carouse with tree huggers. Survivalists lock arms with peaceniks. Tire burners sit down with clean air freaks. And they all do it in the name of a cleaner-burning renewable fuel that is easily made at home in the corner of the garage.

Commercial Production

Bob and I struggled over whether or not we should include a chapter on commercial production in a book on backyard biodiesel. After hours of anguishing, we decided to because accomplished homebrewers think about it. Once you have perfected making fuel to provide motive power for your family, or for your family and friends or for your cooperative, it is impossible not to consider the implications of going commercial. Turning pro, if you like.

Almost everyone we know in the backyard thinks about throwing in the towel on their day job and becoming a commercial biodiesel producer. Since we know you have considered it, here are our thoughts.

Specifications

The fundamental difference between backyard and commercial production is the ability to certify that your fuel conforms to ASTM D6751, the specification for B100 fuel. While it is possible to hit this specification in the backyard, it is expensive to verify that you have done so. Which means going commercial requires some cash. Once you have verified you have done it, you need to share your results with the Environmental Protection Agency (EPA). The least expensive way is to join the National Biodiesel Board as a small producer to gain access to their "human health effects" studies. A small-producer membership costs $2,500—more money than most backyarders have spent on their entire setup.

While the National Biodiesel Board once had a lock on this data, their monopoly expired in 2011, making it possible to become a biodiesel producer without joining the NBB. Since EPA registration is still required, it is not legally clear at this point how you would go about doing that in the absence of the NBB.

Once you have demonstrated a process that can "hit spec," and once you have registered with the EPA, you will want to register with the IRS. That's fun and easy. If there are production credits in place, they will want to visit your facility to ensure that you are a real company, with tanks and pumps and the ability to move liquids around. Then you will need an engineering review that will enable you to register Renewable Identification Numbers (RINS) associated with every gallon you produce. If you are working with waste feedstock, you will be entitled to 1.5 RINS for every gallon you produce, and if you have a certifiable engineering audit, you will be able to sell those RINS, along with your gallons, into the market.

Regulations

Thinking back to the regulations that govern the backyarder, the regulations associated with commercial production are a nightmare. Recall that the government cares more about what you put in your tank than what you put in your mouth.

One regulatory area that differs vastly between the backyard and commercial production is that of "weights and measures." When you are selling fuel to others, you must be able to certify that a gallon is a gallon. No one cares about this in the backyard. Co-ops, collectives and voluntary groups who flaunt this consideration have been closed down across the country. When you are making your own fuel, and dumping it into your own car, close enough is good enough. All you care about is whether the needle on your fuel gauge moves in the right direction (assuming that the sending unit in your tank has not been destroyed by B100). But when you are involved in a commercial transaction, the government cares deeply about how much volume is changing hands.

It's hard for me not to think of the fishmonger who slips their finger onto the scale at the moment of weighing when I think of how this regulation came to pass.

Most backyarders don't bother metering their fuel. If they do, they use a $150 farm meter from Northern Tool. Because such a meter is unable to exclude air, and cannot be calibrated by the user, it is not allowed for commercial use. Meters that comply start at about $2,500 and will be on an approved list maintained by the state.

I have purchased a lot of backyard fuel in my day—all of it illegally—usually when I am out of town on trips. I know how it is entirely possible to pump 14 gallons on air into my Golf, and to drive away feeling full only to find myself empty.

In some ways, the weights-and-measures aspect is analogous to all things associated with commercial production. Things seem to be orders of magnitude more expensive as a commercial producer.

Production and Feedstocks

Compliance is a small part of commercial biodiesel production. The more fiendishly complex task is marrying production to feedstocks. Note that market demand has nothing to do with the equation. Those who build commercial biodiesel plants based on market demand have been bankrupted long ago.

When designing your plant, ignore demand. If you can make a gallon at the same price as petroleum, every drop will be sold before it is even produced. You will need no sales or marketing dollars in your budget.

Where you will need sales expertise and marketing savvy will have nothing to do with selling your product, and everything to do with acquiring feedstocks. Build a plant that you can feed—preferably located on some sort of feedstock anomaly to sustain your commercial endeavor.

There are not a lot of financially successful biodiesel operations in the United States. Those that are have figured out a feedstock anomaly. Perhaps you are on an island, where petroleum is extremely expensive, and discarding used cooking oil is also difficult. Perhaps you are in a cotton patch, where growers are unable to get their oil-rich cottonseeds to the Chicago Board of Trade. Perhaps you are in a state like Oregon, where fuel purchases for State Contract are decided not on the price of petroleum diesel, but rather on the price of biodiesel feedstocks.

One way to be successful as a commercial producer is to build your own anomaly by collecting enough used cooking oil to feed your plant. It's a fiercely competitive thing to do, but it can be done, and it can work profitably—especially if you have sized your production capabilities to match your feedstock acquisition capabilities.

Commercial production is not a license to print money. A lot of investment dollars have been lost, such that raising new capital in the biodiesel space today can be exceedingly difficult. Gone are the days of raising your money by noon—the way Piedmont did.

The commercial biodiesel business in America is wholly dependent on the "policy layer." When production credits are in place, and the Renewable Fuel Standard is demanding a lot of renewable fuel gallons such that RIN prices are high, commercial production can look like genius. When the production credits lapse, and petroleum interests are winning on Capitol Hill such that RIN prices have cratered, commercial production looks like a fool's errand.

While the regulatory aspects of being a commercial producer are daunting compared to backyard, there are also "voluntary" regulations that some commercial producers elect to engage. One is BQ9000, a quality accreditation that is awarded by the National Biodiesel Board. Another is the Roundtable on Sustainable Biomaterials, a certifying scheme that can let your customers know how sustainable your fuel is. Some plants, like Piedmont, regularly commission energy balance studies on their fuel. They are not required, but they are intensely interesting to energy nerds, and they are an expensive voluntary regulation.

Piedmont Biofuels is also a Certified B Corporation, which means we are able to err on the side of our mission rather than on the side of fiduciary responsibility to shareholders. That's important if we want to change the world. And it's a voluntary expense. I once explored this idea in Energy Blog, and our voluntary regulatory costs per gallon, based on 175,000 gallons of production in one year, was a whopping $0.09 per gallon. That's about $15,000 a year just to make a contribution to the industry. Everything gets expensive when you go commercial.

Commercial biodiesel is a roller coaster. Once you are open, you will find yourself sandwiched between global commodity markets over which you have no control. Fats, oils and grease can skyrocket; methanol pricing can mysteriously climb, the entire time that global oil prices decline.

In *Industrial Evolution*, I described commercial production as being somewhat like a mussel, affixed to a rock on the shore. Affix too high and a seabird nails you. A fish nails you when you affix too low. If you get your position just right, you can have a long and productive life, being beaten daily by endless waves and tides.

In biodiesel, sometimes you look like a genius. Sometimes you look like a fool. Better to stay in the backyard and be the envy of your neighbors by consistently beating the price at the pump. It's so much easier to look like a genius when you don't turn pro....

Conclusion

We would like to conclude by saying that making your own fuel is fun and easy, but that is only partially true.

Making fuel takes work. It requires study and attention to detail. It can be a heavily regulated space that requires shrewd negotiation. It can be dangerous, and it requires a great deal of attention to be performed safely.

Part chemistry, part economics, part regulations, part trouble-shooting, making fuel presents a set of diverse challenges. They are easy enough to overcome, if you apply yourself, but they do exist, and each needs to be addressed in order to develop a successful fuel-making operation.

There is no question that biodiesel from used feedstocks makes a tremendous amount of environmental sense. And there is also no question that it can be hard to scale up beyond personal fuel needs.

There are a lot of dead bodies along the side of the biodiesel road. Most of them are commercial operators destroyed by im-perfect financial models. Some are community-scale projects killed by personality clashes or unrealistic expectations. Some are co-ops, where the promise of motive power from waste never rose above the power of the status quo.

Biodiesel is a difficult space. It can work well in the back-yard—with proper planning and attention. And biodiesel is addictive. When it is working, you have an amazing feeling of freedom. Nothing is finer than exiting the petroleum grid and driving down the road on fuel you have fashioned for yourself.

We think you should go for it. Pay attention, and give it a try. Be safe, stay legal and enjoy the ride.

We certainly have....

Index

About the Authors

Lyle Estill is a biodiesel enthusiast. He's also a serial entrepreneur who is actively involved in mission-driven social enterprise. When he is not out in the world speaking, writing or teaching, you can find him at home—hidden in the woods of Chatham County, North Carolina. There he grows vegetables and citrus, forages for wild edibles and harvests wild game.

His work in sustainable fuel has spilled over into green building, co-housing, solar, wind and hydro enterprises. He is the co-founder of Slow Money North Carolina, and Abundance North Carolina, and he is actively involved in local currency and time-banking endeavors.

Part activist, part recluse, part community organizer, Estill is best known in Chatham County for his astonishing head of hair.

Bob Armantrout spent the last 14 years working on resilience—personally and professionally. In 2001 he bought a VW Beetle TDI and ran it on 100% biodiesel because using a locally available waste resource to offset fossil fuel use was just too easy. His day job on Maui was driving a 100% biodiesel-fueled F450 recycling truck for Maui Recycling  Service. He then became Operations Manager for Pacific Biodiesel's plants on Maui and O'ahu where he learned more about processing grease trap waste than making biodiesel.

Bob moved back to the mainland in 2005, where he took on the responsibility of managing a small biodiesel plant in Berthoud, Colorado. He became a GC Jedi, and met Lyle from Piedmont Biofuels who was in town for a Solar Energy International gig in Carbondale. He and his wife, Camille, decided to move east and throw in with the folks at Piedmont.

If you have enjoyed *Backyard Biodiesel*, you might also enjoy other

BOOKS TO BUILD A NEW SOCIETY

Our books provide positive solutions for people who
want to make a difference. We specialize in:

**Food & Gardening ◆ Resilience ◆ Sustainable Building
Climate Change ◆ Energy ◆ Health & Wellness
Sustainable Living ◆ Environment & Economy
Progressive Leadership ◆ Community
Educational & Parenting Resources**

New Society Publishers
ENVIRONMENTAL BENEFITS STATEMENT

New Society Publishers has chosen to produce this book on recycled paper made
with 100% post consumer waste, processed chlorine free, and old growth free.

For every 5,000 books printed, New Society saves the following resources:[1]

16	Trees
1,414	Pounds of Solid Waste
1,556	Gallons of Water
2,030	Kilowatt Hours of Electricity
2,571	Pounds of Greenhouse Gases
11	Pounds of HAPs, VOCs, and AOX Combined
4	Cubic Yards of Landfill Space

[1]Environmental benefits are calculated based on research done by the Environmental Defense Fund and
other members of the Paper Task Force who study the environmental impacts of the paper industry.

For a full list of NSP's titles, please call 1-800-567-6772 or check out our web site at:
www.newsociety.com